Molecular Biology
Biochemistry and Biophysics
20

Max A. Lauffer

Entropy-Driven Processes in Biology

Polymerization of Tobacco Mosaic Virus Protein and Similar Reactions

With 90 Figures

Springer-Verlag Berlin Heidelberg GmbH 1975

Professor Max A. Lauffer, Ph. D.
Faculty of Arts and Sciences
University of Pittsburgh
Department of Biophysics and Microbiology
Pittsburgh, PA 15260/USA

ISBN 978-3-642-80871-5 ISBN 978-3-642-80869-2 (eBook)
DOI 10.1007/978-3-642-80869-2

Softcover reprint of the hardcover 1st edition 1975

Library of Congress Cataloging in Publication Data. Lauffer, Max Augustus, 1914–. Entropy-driven processes in biology. (Molecular biology, biochemistry, and biophysics, v. 20). Bibliography: p. 1. Tobacco mosaic virus. 2. Entropy. 3. Biological chemistry. I. Title. II. Series. [DNLM: 1. Polymers. 2. Proteins. 3. Tobacco mosaic virus. 4. Water. WL M195T no. 20 / SB736 L373e]. QR 467.L38.576'.6483.74-18267

Preface

The purpose of this monograph is to bring together under one cover results of research on phenomena drawn from the fields of chemistry, biochemistry, biophysics, virology, and cell biology. The processes and reactions considered have one important feature in common: they are endothermic and, therefore, entropy-driven. They are, in the main, reversible reactions leading to the formation of large structures, some of which play critical roles in life processes.

If one thinks only of the subunits and of the structures they form upon polymerization, it seems to be a contradiction that such reactions can be driven by an increase in entropy; entropy is a measure of disorder. The increase in entropy must come from some other source, usually from the release of something coincidental to polymerization. That something has been shown to be water for the case of the polymerization of tobacco mosaic virus protein. Because of the remarkable similarity of the other processes to this one, it is a permissable inference that the release of water is the source of the entropy increase and therefore the driving force for all of them.

The reactions and processes brought together in this book are still the subjects of active research. Many of the detailed interpretations presented here must be regarded as tentative, subject to modification as new information becomes available. However, the main characteristic of each reaction or process, its endothermic or entropy-driven nature, is well established in all but one or two instances.

The structures dealt with in the following pages are unusual in two important respects. First, they are very readily reversed by minor changes in environment, for example, small changes in temperature, in pressure, in *pH*, in concentration of chemicals, etc. The significance or the potential significance of processes with this characteristic for cell division, muscular contraction, etc. must not be overlooked. Secondly, these structures are not held together primarily by chemical bonds in the ordinary sense. The structures are stable under some conditions of temperature, *pH*, etc. primarily because interaction with water would be thermodynamically disadvantageous if the structures were to dissociate. Under other conditions, interaction of dissociated units with water is thermodynamically favorable; the structures then dissociate. This key role of water gives us new insight into the biological importance of water.

It is appropriate, therefore, to dedicate this book to the memory of R.A. GORTNER, for more than a quarter of a century the Chief of the Division of Agricultural Biochemistry at the University of Minnesota. Professor GORTNER was the author's major advisor during his time of graduate study. He recognized the importance of water in living systems far in advance of his time. His enthusiasm undoubtedly encouraged the author's interest in the widely scattered phenomena treated in this book.

Pittsburgh, Autumn 1974 M.A. LAUFFER

Acknowledgment

Much of the work reported herein was done in the author's laboratory and was supported for many years by U.S. Public Health Service Grant, GM 10403.

Contents

Chapter 1

The Nature of Entropy-Driven Processes

Division of fertilized eggs, protoplasmic streaming, and the formation of pseudopodia are all processes which have at least one thing in common; they are, in many widely different organisms, accompanied by the transformation of protoplasm from a sol to a gel, an equilibrium reaction in which the structured state is favored by an increase in temperature. Several well-known biochemical reactions, such as polymerization of tobacco mosaic virus protein, polymerization of sickle-cell anemia hemoglobin and formation of collagen fibers from soluble collagen, are also equilibrium processes in which the highly structured state is favored by an increase in temperature. Normally, the reverse is true; heat dissociates. Therefore, these biological and biochemical processes are unusual. They and others like them are the subject matter of this book and are discussed in detail in subsequent chapters. Several have been reviewed briefly by the author on a previous occasion (1).

Spontaneous Processes

Spontaneous reactions or processes are invariably accompanied by a decrease in Gibbs free energy, F. This statement can be represented symbolically by saying ΔF is negative. F is defined as $H - TS$, where H means the heat content or the enthalpy, T means the absolute temperature and S means the entropy. For a reaction or a process which occurs at constant temperature, an isothermal process,

$$\Delta F = \Delta H - T\Delta S. \quad (1)$$

ΔF is the useful work which could be done as a result of the process if it were carried out reversibly in an appropriate apparatus such as, for example, a galvanic cell. It is equal to the heat absorbed at constant pressure, ΔH, minus T times the increase in entropy, ΔS. For most spontaneous (non-reversible) reactions or processes, ΔF is negative because ΔH is negative, i.e., because the enthalpy of the products is less than the enthalpy of the reactants, or, to state it differently, because heat is given up to the surroundings when the process or reaction occurs. However, there are exceptions. One of the best known is the melting of ice at temperatures above melting point. This is a spontaneous process; otherwise it would not occur. Therefore, ΔF is negative. But heat is absorbed from the surroundings; for the system itself, ice-water, there is an increase in heat content or in enthalpy; ΔH is positive. When ΔH is positive, ΔF can be negative only when ΔS is positive and large enough so that $T\Delta S$ exceeds ΔH in magnitude. Such a transformation proceeds, then, because of the increase in entropy. This is what is meant by an entropy-driven process. The example chosen is also an endothermic process because heat is absorbed by the system during the spontaneous change. All spontaneous endothermic processes must of necessity be entropy-driven.

There are many examples, in addition to the melting of ice at temperatures above freezing point, of transformations or reactions which proceed spontaneously, even though accompanied by a positive ΔH. One is the reaction between silver and mercurous chloride, both in the solid state, to give solid silver chloride and liquid mercury. Another, frequently used as a lecture demonstration (2), is the reaction between one mole of crystalline hydrated barium hydroxide and two moles of crystalline ammonium thiocyanate to yield one mole of barium thiocyanate, two moles of ammonia and ten moles of water. This spontaneous reaction absorbs sufficient heat to freeze several grams of water outside the reaction flask. In a sense, any chemical transformation which involves an equilibrium can be an example. The reason is that if one starts with an equilibrium mixture of reactants and products and then adds a reactant or a product, the equilibrium will be re-established by the reaction proceeding spontaneously in one direction or the other, depending upon what was added. If the process is exothermic in one direction, it must of necessity be endothermic in the other, and therefore entropy-driven.

From thermodynamics one learns that

$$-RT \ln K = \Delta F^{\circ}, \quad (2)$$

[1]

where K means the numerical value of the equilibrium constant. Here, ΔF° means the standard increase in free energy, the free energy increase for transforming reactants into products when all reactants and all products have, for reactions in solution, concentrations of 1 mole per liter or, better, unit activity. The standard free energy change must not be confused with the actual free energy change; ΔF° is a special case applicable only to standard conditions of unit activity for all participants. It is ΔF, the free energy under the actual conditions of a reaction, which determines whether or not a process is spontaneous.

By combining Eqs. (1) and (2), one obtains:

$$\ln K = \Delta S^{\circ}/R - (\Delta H^{\circ}/R)(1/T). \quad (3)$$

This equation is correct only for a process which takes place at constant temperature. ΔS° is the entropy increase for the standard condition of unit activity of all participants. ΔH° is the enthalpy for standard conditions of unit activity. ΔH does not usually vary much on concentration with biological materials such as protein. If ΔH° and ΔS° both remain unchanged when the process is carried out isothermally at several different temperatures, then a straight line is obtained, as illustrated in Fig. 1, with a slope of $-\Delta H^{\circ}/R$, when $\ln K$ is plotted versus $1/T$. Values for ΔH° can be obtained by the use of this equation from measured values of the equilibrium constant, K, at different temperatures. For historical reasons, enthalpies obtained in this manner are called van't Hoff enthalpies.

[1] The mathematically vigilant will marvel at the logarithm of K, usually not a dimensionless quantity. Actually, the derivation of Eq. (2) found in standard textbooks of physical chemistry shows that K appearing there is an abbreviation for the equilibrium constant multiplied by the reciprocal of its dimensions. Since the numerical value of K depends upon the manner of expressing concentration, so does the value of ΔF°. It is thus necessary to specify the standard state (e.g., unit molarity, unit mole fraction, etc.). Except when otherwise specified, ΔF° and K will be reported on the basis of unit molarity.

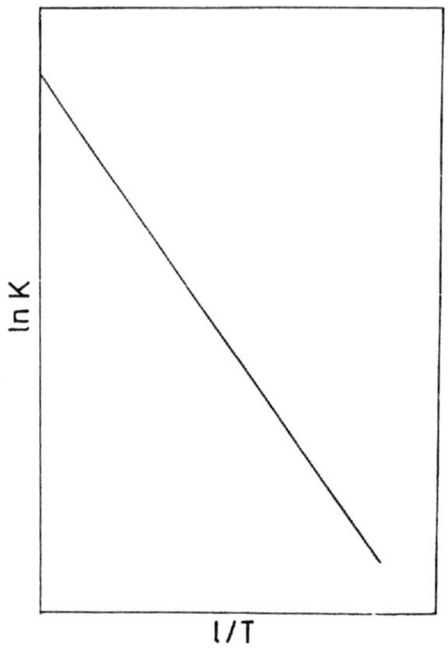

Fig. 1. Theoretical plot of the logarithm of the equilibrium constant, ln K, against the reciprocal of the absolute temperature, $1/T$, in accordance with Eq. (3)

If one imagines changing reactants at temperature T_1 to products at a higher temperature, T_2, there are two ways that this can be done. The first is to perform the reaction at the lower temperature and then raise the temperature of the products. The total heat change will be $\Delta H_1 + \Delta T C_{pp}$, where C_{pp} is the heat capacity at constant pressure of the products. As an alternative, the temperature of the reactants can first be raised from T_1 to T_2 and then the reactants can be transformed into products. The total heat change at constant pressure is then $\Delta H_2 + \Delta T C_{pr}$, where C_{pr} is the heat capacity of the reactants. According to Hess's Law, the total heat change should be the same in both cases.

$$\Delta H_1 + \Delta T C_{pp} = \Delta H_2 + \Delta T C_{pr}. \tag{4}$$

Therefore,

$$(\Delta H_2 - \Delta H_1)/\Delta T = C_{pp} - C_{pr} \tag{5}$$

or,

$$\partial \Delta H/\partial T = \Delta C_p \tag{6}$$

where ΔC_p is the difference in heat capacity at constant pressure between products and reactants. This is known as the Kirchoff relationship. Obviously, only when ΔC_p is zero can ΔH be independent of temperature.

The Nature of Entropy

Thermodynamically, entropy is defined by the equation, $dS = dH/T$ or, for an isothermal process, $\Delta S = \Delta H/T$, where ΔH is the heat absorbed per mole when the transformation is carried out reversibly. This is obvious from Eq. (1) for an isothermal process, because, when the transformation is carried out reversibly, ΔF is zero. For reactions in solution,

$$\Delta F = \Delta F^{\circ} + RT\Sigma\nu_i \ln a_i, \tag{7}$$

where a_i is the activity of the ith constituent and ν_i is the number of moles of the ith constituent formed; ν_i is negative when a constituent is consumed. Therefore,

$$\Delta H - T\Delta S = \Delta H^{\circ} - T\Delta S^{\circ} + RT\Sigma\nu_i \ln a_i. \tag{8}$$

In this equation, ΔH will equal ΔH° when ΔH is independent of the activities of the constituents, which will be the case when the heats of dilution of all constituents are negligible. Therefore,

$$\Delta S = \Delta S^{\circ} - R\Sigma\nu_i \ln a_i. \tag{9}$$

Thus, there are two contributions to ΔS, ΔS° and $- R\Sigma\nu_i \ln a_i$. Since the activity term is different for a solution in equilibrium and for one not in equilibrium, ΔS will also be different in the two cases.

Consider a polymerization reaction in aqueous solution in which a protein monomer with two reaction sites, A and B, one on each end, reacts with a polymer, to form a longer polymer. This reaction can be symbolized as:

$$APB + AP_nB = AP_{n+1}B \tag{10}$$

ΔS° for this reaction will correspond to the entropy change when all components have unit activity, or, if the solution is ideal, a concentration of 1 mole per liter. When $\overline{S}$ means entropy per mole,

$$\Delta\overline{S}^{\circ} = \overline{S}^{\circ}_{P_{n+1}} - \overline{S}^{\circ}_{P} - \overline{S}^{\circ}_{Pn} \tag{11}$$

The three $\overline{S}^{\circ}$ terms represent the entropy per mole of the three components when each has a concentration of 1 mole per liter. But the molar entropy of each is greater than it would be if that solution were more concentrated. Thus, it is composed of two terms, the intrinsic entropy or the unitary molar entropy, $\overline{S}_u$, and the entropy of mixing, $-R \ln x$, where x is the mole fraction. The term, $-R \ln x$, is called the cratic contribution to the entropy. Thus,

$$\overline{S}^{\circ} = \overline{S}_u - R \ln x_{\circ}, \tag{12}$$

where $x_{\circ}$ is the mole fraction corresponding to a concentration of 1 mole per liter. For aqueous solutions the numerical value of $x_{\circ}$ is 1/55.6, since a liter contains 55.6 moles of water. Therefore,

$$\Delta\overline{S}^{\circ} = \overline{S}_{uP_{n+1}} - R \ln x_{\circ} - \overline{S}_{uP} + R \ln x_{\circ} - \overline{S}_{uP_n} + R \ln x_{\circ},$$

or

$$\Delta\overline{S}^{\circ} = \Delta\overline{S}_u + R \ \ln x_{\circ}. \tag{13}$$

The standard molar entropy change, $\Delta\overline{S}^{\circ}$, is what one evaluates from equilibrium studies. However, the unitary molar entropy change, $\Delta\overline{S}_u$, is frequently the term of interest, because it is free of the accidental term associated with concentration and is a true measure of other contributions to the entropy change, such as interaction with the solvent.

Since $x_{\circ}$ is 1/55.6, $- R \ln x_{\circ}$ is -1.98 cal/mole.deg times -4.02 or 7.98.

$$\Delta\overline{S}_u = \Delta\overline{S}^{\circ} + 7.98 \text{ e.u.} \tag{14}$$

For some processes, such as changing concentration of a solution when ΔH of dilution is zero, ΔF is $-T\Delta S$. For other processes, such as those in an electrolytic cell, ΔF equals minus the electrical work or minus the potential times the amount of electricity. This is the product of an intensity factor (potential) and a capacity factor (amount of electricity). By analogy, $T\Delta S$

can be considered a product of the intensity factor, T, and a capacity factor, ΔS.

From statistical mechanics, one learns that entropy is a measure of disorder. When a process is carried out in which a highly organized substance, such as a crystal, disperses into many smaller particles, such as molecules in solution, there is an increase in entropy. It is, therefore, not surprising to find entropy-driven spontaneous processes which involve the formation of many small particles (disordered) from a single large particle (ordered). However, in biological and biochemical processes, one frequently encounters endothermic and, therefore, entropy-driven reactions in the building of large structures from subunits. Among them are those mentioned in the first paragraph. This is surprising because the examples are such in which complex, highly organized structures are formed from smaller dispersed building blocks. They are all endothermic processes, recognized by the fact that structures are formed at higher temperatures and dissociated at lower temperatures. Therefore, they must be accompanied by an increase in entropy. Taken at face value, this is a contradiction.

Theory of Entropy-Driven Processes

Endothermic and, therefore, entropy-driven reactions which lead to the formation of structured material have been known for a long time. An endothermic sol-gel transformation was discovered in 1924 by SZEGVARI (3). If nitrocellulose is dissolved in amyl acetate and benzine is then added, a viscous sol is obtained at room temperature. When the temperature is increased, a rigid gel forms; when it is lowered fluidity returns. HEYMAN (4) reported in 1935 that colloidal aqueous solutions of methylcellulose undergo a sol-gel transformation with gel formation being favored at higher temperatures. This process was discussed by FREUNDLICH (5) in 1937. Both of these phenomena were explained on the assumption that the colloidal particles or macromolecules bind the principal solvent in the sol state. When the temperature is increased, the macromolecules join together to form a structured network with the simultaneous release of the bound solvent molecules. This release of bound solvent provides the increase in entropy necessary to drive the structure-forming reaction. Thus, a theory has been available for a long time to explain endothermic or entropy-driven structure-forming processes in biochemistry and biology. Even though the formation of structure involves a decrease in entropy, according to this theory this decrease is over-balanced by an increase in entropy coming from the release of bound solvent molecules. In 1965, STEVENS and LAUFFER (6) demonstrated that water molecules actually are released when tobacco mosaic virus protein polymerizes, thereby furnishing the first experimental evidence to verify this theory. This experiment and others like it will be discussed in detail in Chapter 2.

Another observation was made by HEYMAN (4) in 1935. His colloidal aqueous solutions of methylcellulose showed an increase in volume upon gelation. Similar evidence for an increase in volume when tobacco mosaic virus protein polymerizes was obtained by STEVENS and LAUFFER (6).

In accordance with the Le Chatelier principle, if there is an increase in volume in a reaction which leads to the formation of a structure, then the structure-forming process can be reversed by the application of hydrostatic pressure. From thermodynamics, at constant temperatures

$$dF/dP = V, \text{ or } d\Delta\overline{F}/dP = \Delta\overline{V}. \tag{15}$$

From this, one can write,

$$d\Delta\overline{F}^{o}_{o}/dP = -RT\, d\, \ln K/dP = \Delta\overline{V} \tag{16}$$

Upon integration one obtains,

$$\ln K = \ln K_o - (P - P_o)\, \Delta\overline{V}/RT. \tag{17}$$

In this equation K_o means the equilibrium constant at P_o, the atmospheric pressure. Thus, if one plots $\ln K$ against $(P - P_o)$, one should obtain a straight line with a negative slope equal to $\Delta\overline{V}/RT$. This is illustrated diagramatically in Fig. 2. Moreover, since the equilibrium constant for polymerization or for structure formation increases when temperature is increased for endothermic or entropy-driven processes, $\ln K_o$ will be higher at higher temperatures. Since one can work with biological materials only at temperatures above $273^o\ K$ and below about $310^o\ K$, the RT factor will be nearly constant. Thus, if $\Delta\overline{V}$ does not vary much with temperature, one should expect straight lines with nearly equal slopes at the different temperatures. This is also illustrated diagramatically in Fig. 2. In summary, a volume change inherent in a structure-forming process can be detected either by direct measurement or by studying the effect of hydrostatic pressure on the equilibrium.

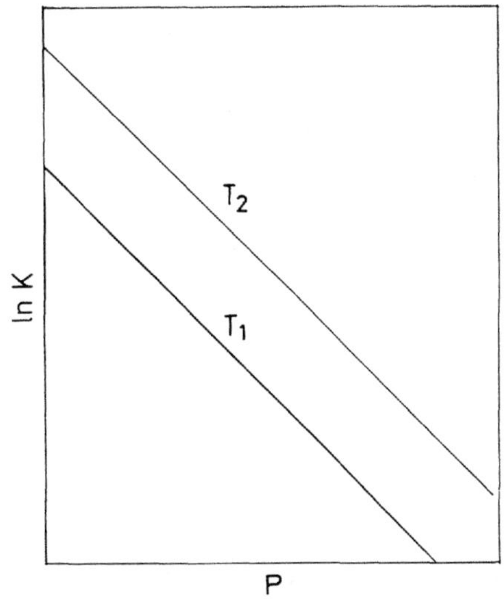

Fig. 2. Theoretical plot of the logarithm of the equilibrium constant, ln K, against the pressure, P, in accordance with Eq. (17). Theoretical plots are shown for two temperatures, T_2 and T_1, T_2 being the higher.

By rewriting Eq. (3) with subscripts, o, and substituting for ln K_o in Eq. (17), Eq. (18) can be derived; it is presented in both logarithmic and exponential forms.

$$\ln K = \Delta S_o^o/R - \Delta H_o^o/RT - (P - P_o)\ \Delta\overline{V}/RT$$

$$K = e^{-\Delta H_o^o/RT}\ e^{-\ (P - P_o)\Delta\overline{V}/RT}\ e^{\Delta S_o^o/R} \qquad (18)$$

ΔS_o^o and ΔH_o^o refer to the values of these parameters at atmospheric pressure, P_o. Eq. (18) is useful for showing simultaneously the relationship between an equilibrium constant and both temperature and pressure.

Entropic Unions

There remains the necessity for discussing in this introductory chapter the nature of the "bonds" involved in the entropy-driven processes in biological systems. Most of the structure-forming units with which we will be concerned are proteins. The surface of the protein molecule can contain a variety of chemical groups, polar groups capable of forming hydrogen bonds with water, ionizable groups capable of forming positive or negative ions in a medium of high dielectric constant such as water, most of which will be ionized in water at the *pH* values normally found in living systems, and non-polar organic residues. Each kind must interact with water when in contact with water, as it must be when the protein molecule is in its unpolymerized state and its surface is surrounded by water. It is reasonable to suppose that for each specific group on the surface of the protein molecule, there are specific ΔH and ΔS values for water interaction. Whether the free energy for interaction with water of that group is positive or negative depends on the temperature. The total free energy for water interaction of an element of surface will be determined by the contributions of all the various groups on that surface, allowing for possible interaction among them. When protein molecules polymerize, portions of the surface of each molecule are removed from contact with water and come in contact with other protein molecules. Polymerization thus involves the transfer of certain groups on the surface of the protein molecule from an aqueous to an essentially organic environment. When the total free energy of the interaction of that element of the surface with the organic environment is less than the free energy of the interaction with water, there will be a decrease in free energy when the protein molecules polymerize. Under such circumstances, the equilibrium between protein monomer and polymer will be shifted in the direction of polymerization. At a different temperature, the free energy for water interaction might be less than the free energy for interaction with an organic environment. At that temperature depolymerization will occur.

What is needed to explain endothermic or entropy-driven polymerization is some model substance for which the free energy change for transfer from an aqueous environment to an organic environment is negative at room temperature and positive at temperatures just above freezing point.

KAUZMANN (7), stimulated by the studies of FRANK and EVANS (8), showed that the properties of small hydrocarbon molecules can shed much light on this question. When one mole of methane is transferred from a dilute aqueous solution to a solution of the same mole fraction in ether, $\Delta\overline{H}$ and $\Delta\overline{S}_u$ are +2400 cal and +19 cal/deg or +19 e.u. From the enthalpy and entropy one can calculate that, at 25°C, $\Delta\overline{F}_u$ is -3300 cal/mole. At this temperature methane would move spontaneously from an aqueous solution to an ether solution at the same mole fraction.

Bond energies are usually defined as the average amount of heat required to break a bond of a given type or evolved when the bond is formed. These energies range between 50 and 150 kcal/mole for the bonds frequently encountered in organic compounds. When a gas like methane liquifies, heat to the extent of about 2 kcal/mole is evolved. Thus, there are weak forces of attraction, van der Waals forces, or "bonds" between methane molecules in the liquid state. Similar "bonds" exist in solutions between methane molecules and organic solvent molecules. Since heat is evolved when methane is transferred from ether to water, the "bonds" between methane and water are stronger than between methane and ether. Nevertheless, at 25°C methane moves spontaneously from an aqueous solution to an ether solution at the same mole fraction. The driving force is the very large increase in entropy. Thus, this is an entropy-driven process.

The source of this entropy increase is known. When organic molecules like methane and some of the inert gases dissolve, they induce water to assume a more structured form. A review of current ideas concerning the nature of such structure formation has been presented by BEN-NAIM (9). No complete theory is yet available. When the methane is removed, these induced structures disappear and the water reverts to the normal configuration of liquid water. There is thus a decrease in entropy when methane is dissolved in water and an increase in entropy when it is removed. It is reasonable to suppose that, if a methyl group is a side chain of an amino acid residue on the surface of a protein, similar water structures will be induced when that methyl group is in contact with water. This structure will break, providing an increase in entropy, when the surface is removed from water by polymerization.

On transferring an element of protein surface from an aqueous to an organic environment, one destroys the interaction between that element and water, with its characteristic ΔH and ΔS and creates interaction with the organic medium with its characteristic ΔH and ΔS. Thus the total free energy change for the transfer depends upon the difference between the ΔH values and the ΔS values for the two interactions. If the net entropy increase multiplied by the absolute temperature exceeds the net enthalpy increase, there will be a decrease in free energy for the transfer. The thermodynamically stable state will be the one in which the surface is in an organic environment, such as in polymerized protein. It must be emphasized that the thermodynamic stability is not primarily related to the interaction of the protein suface with an organic medium, that is, the adjacent surface of another protein molecule in the polymer chain, but rather to the difference between the interaction with an aqueous medium and an organic medium. This differs from the situation which exists in a structure like that of a diamond, where the stability is related entirely to the interaction of carbon atoms with neighboring carbon atoms. One normally thinks of a chemical bond as being something like the carbon-carbon interaction in a diamond.

Even though the factors responsible for the stability of structures formed by an entropy-driven process are much more complex than that involved generally in chemical bonds, the name "hydrophobic bond" has been applied to this situation. TANFORD (10) pointed out that the word "hydrophobic" is a misnomer, for it implies that the dissolved substance is averse to water whereas, in fact, it is the water that is averse to the dissolved substance. LAUFFER (11) pointed out that the word "bond" is misleading because there need not be any force at all between the protein surface and its adjacent protein surfaces in the polymer structure. Even if there are such forces, they contribute only partially, perhaps only minimally, to the thermodynamic stability. LAUFFER (11), therefore, maintained that"entropic union" describes the actual situation better than "hydrophobic bond." In the subsequent discussion, we will use "entropic union" rather than "hydrophobic bond," which is frequently used by others to mean the same thing.

According to data assembled by KAUZMANN (7), there is also an increase in volume of 22.7 ml/mole when methane is transferred from water to hexane. We assume that it would be about the same for the transfer to ether. This volume increase (for the whole system, aqueous solution and organic solution) originates from a closer packing of water molecules in the more structured form than in ordinary water. This volume increase makes the small organic molecules even more appropriate as a model for understanding the nature of entropy-driven reactions in biological systems because, as was already mentioned, there is also an increase in volume in such systems.

Many other processes, some of great biological significance, can at least partially be attributed to entropic interactions with water. Among them are conformational changes in proteins. Some of these are so subtle that it would be unrealistic, as was pointed out by BEN-NAIM (9), to view the mechanism of change as the complete removal of hydrocarbon side chains from contact with water. Changes in entropic interaction are more likely to be related to overlapping of the zones of water interaction when side chain are brought closer together.

There is a category of reaction which exhibits enthalpy-entropy compensation. LUMRY and RAJENDER (12) reviewed such phenomena comprehensively and advanced the thesis that the explanation common to all lies in water-interaction. Any reaction in a living system, if it is to be reversible, must have an actual free energy change, ΔF, of zero somewhere within the relatively narrow temperature range within which life processes are possible. The standard free energy change, ΔF°, can fall within a slightly broader range. From Eq. (1) it follows that ΔF° will be zero for a given process only at a particular temperature equal to $\Delta H^{\circ}/\Delta S^{\circ}$, and this temperature must not be far removed from 300°K if the process is to be reversible in a biological system. It has been observed for a large number of processes taking place in aqueous solution that if ΔH° is changed in some way, there is a compensating change in ΔS°. This is what is meant by enthalpy-entropy compensation. In the simplest case, illustrated by the ionization of primary alkyl ammonium compounds in aqueous solution, there is an accurate linear relationship between ΔH° and ΔS° when both of these parameters are varied by changing the length and configuration of the alkyl chain. The slope of the straight line has the dimensions of absolute temperature and is called the compensation temperature, T_c. The value of T_c is 308°K. The explanation is that changes in the enthalpy and entropy of ionization depend primarily upon changes in the interaction of the ions with water. Many other processes yield similar straight lines with values of compensation temperature ranging from 250° to 315°K but, in many instances, only after the water interaction contribution is isolated by one means or another from other changes in ΔH° and ΔS°. Among the examples considered by LUMRY and RAJENDER are the solubility of argon in water-methanol and in water-ethanol solvents, the binding of N-acetyl-L-tryptophan and other molecules to α-chymotrypsin, the transfer of non-electrolytes and ions from water to heavy water, the effect of ethanol on the unfolding of ribonuclease, the solution of short-chain alcohols and alkanes in water and the binding of SCN^- and various ligands to hemoglobins. LUMRY and RAJENDER make the interesting suggestion that protein behavior provides a good experimental basis for investigating the properties of water solutions.

The immediately preceeding discussion illustrates that interaction with water applies to a broader range of phenomena than the formation of entropic unions in entropy-driven polymerizations. Conversely, there are also some entropy-driven synthetic processes which seem not to depend upon interaction with solvent. An example comes from the studies of STEVENS and FELSENFELD (13). When a homopolymer of riboadenylic acid (poly A) is mixed in a one to one ratio with the homopolymer of polyribouridylic acid (poly U) at room temperature in aqueous solutions, a two-stranded mixed polymer, poly (A·U) is formed. At higher temperatures in solutions containing more than 0.2 M sodium ion, two poly (A·U) 's are converted into the triple-stranded polymer, poly (U·A·U), and free poly A. This synthetic reaction is driven by an increase in entropy, but the entropy increase can be attributed to the uncoiling of the poly A strand released.

Other Modes of Water Binding

Water can also be bound on an element of a protein surface by electrostriction at either positive or negative groups such as NH_3^+ or COO^-. If a pair of charged amino and carboxyl groups is transferred from an aqueous medium, in which they will be charged, to an organic medium of low dielectric constant, where they must both be uncharged, a proton will migrate from the NH_3^+ to the carboxyl group. It is possible to estimate the thermodynamic parameters of this process by considering the reaction; acetate ion + ethylammonium ion yields acetic acid + ethyl amine. The enthalpy change and the standard entropy change (same as the unitary entropy change in this case) are approximately 13,400 cal and 17.2 e.u., respectively (11). $\Delta\bar{H}$ and $\Delta\bar{S}_u$ were estimated (11) from data taken from the International Critical Tables for the transfer of un-ionized acetic and ethyl amine from an ideal aqueous solution to a solution in ether with the same mole fraction. For acetic acid, $\Delta\bar{H}$ is -565 cal/mole and $\Delta\bar{S}_u$ is 0. For ethyl amine, both ΔH and $\Delta\bar{S}_u$ are approximately 0. The silmultaneous de-ionization and transfer of one mole each of acetate ion and ethylammonium ion from water to an organic solvent without change in mole fractions involves a value of $\Delta\bar{H}$ of 12,835 cal/mole, and a value of $\Delta\bar{S}_u$ of +17.2 e.u. From data given by

EDSALL (15), the volume increase for the de-ionization can be estimated to be +16.3 ml/mole. This volume increase comes about largely from the release of water bound by electrostriction, for in this case, water dipoles are oriented around the charges and the water molecules are closer together than in bulk water.

There are also groups on an element of protein surface which are capable of forming hydrogen bonds when in contact with water. An example is the OH group of certain amino acid residues. From data assembled by COHN and EDSALL, (16) one can calculate that the unitary free energy change, $\Delta\overline{F}_u$, for transferring alanine from water to ethyl alcohol at 25°C is +3920 cal/mole, whereas that for transferring serine is +4620 cal/mole. The difference, +700 cal/mole, can be attributed to the disruption of hydrogen bonds between the OH group and water, since the only difference between these two amino acids is the substitution of an OH for an H. Furthermore, the apparent molar volume of alanine in water is 60.6 ml and that of serine almost the same, 60.8 ml. Yet serine has a molecular weight of 105.09 compared to alanine with 89.09. Thus, there must be some contraction of water molecules in the neighborhood of the OH group on the serine, and one would expect a volume increase when an OH group is transferred from water to an organic medium.

In the example just discussed there is a negative free energy change to be attributed to the formation of hydrogen bonds between the OH group and water when the transfer is made from ethyl alcohol to water. This is an example of the interaction of a group with water in which the negative enthalpy exceeds the positive contribution to the free energy derived from the decrease in entropy. In other words, this is an enthalpic process involving water interaction. Thus, water interaction is not limited to entropy-driven processes. Indeed, it is widely recognized that such enthalpic processes as the stabilization of structures by internal hydrogen bonds depend upon the difference between the internal hydrogen bond energy and the bond energies for the interaction of the separate groups with water. Nevertheless, the phenomena discussed in this monograph have to deal mainly with entropic interaction involving water.

Complex Surface Elements

The protein surface is complex, containing many different kinds of side chain. The net effect of removing the protein surface must depend upon the effects of removing all of the elements capable of interacting with water. When an element of surface on a protein monomer is transferred from an aqueous medium, as encountered when the protein is dissolved in water in the monomeric state, to an organic medium, as encountered after the protein polymerizes, all of the interactions between that surface and water will be disrupted. It seems that, no mater what the nature of the interaction with water, this results in an increase in volume. The ΔH and the ΔS will be the net, allowing for possible interactions among them, of the contributions from the various kinds of group. Whether or not the element of surface considered as a unit will transfer spontaneously from an aqueous to an organic environment depends only upon whether the net of the contributions of all groups to the free energy change is negative. It is possible, therefore, for some side chain to be carried from an aqueous environment to an organic environment, even though the contribution made by that side chain alone to the free energy is positive. It will be dragged along by the contributions of neighboring side chains with negative contributions to ΔF great enough to overcome it. One can easily imagine combinations of side chains on proteins which will produce ΔH and ΔS values for transfer comparable to those observed experimentally in the polymerization of protein.

Conformational Change

Finally, a word must be said about conformation. When a protein molecule is exposed on all sides to water, one would expect it to assume a conformation which presents the maximum possible number of hydrophilic side chains and the minimum possible number of hydrophobic side chains to water. The primary structure or amino acid sequence in the protein might make it impossible for all hydrophobic side chains to be buried in the interior. After the protein has become polymerized and portions of its surface are in contact with other protein molecules, that is, with organic matter, the tendency for hydrophilic side chains to remain on the surface should no longer exist. Thus, there is a real possibility that, coincidental to polymerization, there may be some change in conformation. One of the experimental tools used to study conformation is optical rotatory dispersion. Changes in optical rotatory dispersion have been observed when tobacco mosaic virus protein polymerizes (17, 18, 19). The most conspicuous feature is the reversible change in the depth of the trough of the negative Cotton effect at a wavelength of 232 mμ, shown in Fig.3. A Cotton effect with a trough at such a wavelength is usually interpreted as evidence for the presence of α-helices in the protein structure. Furthermore, tobacco mosaic virus protein in the unpolymerized form exhibits a circular dichroism curve of the sort regarded as evidence for α-helices (20). When the protein is polymerized by raising the temperature, the depth of the trough of the negative Cotton effect decreases. Also when the protein is polymerized by lowering the *pH*, the circular dichroism spectrum shifts. Taken at face value, these results would indicate a change in the degree of helicity upon polymerization. However, URRY ET AL. (21) have shown that circular dichroism

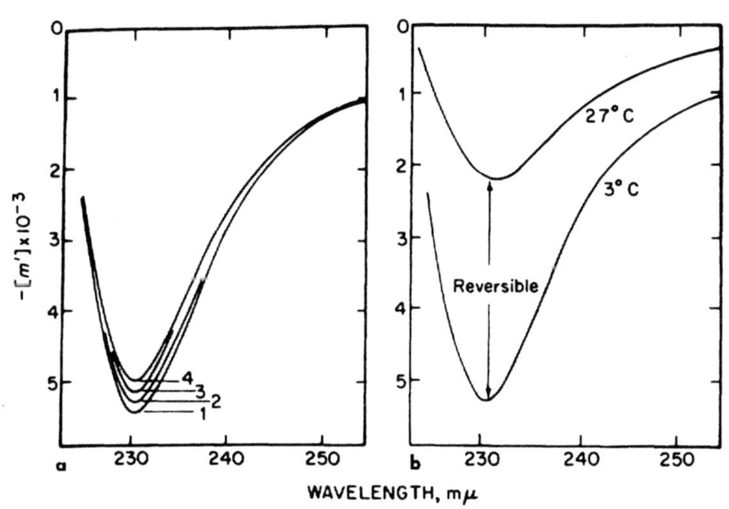

Fig. 3. Optical rotatory dispersion of TMVP. Influence of temperature. (a) 0.01 μ phosphate buffer, *pH*=7.0, concentration-1.6 milligrams per milliliter, 1-millimeter path length. (1) 3°C, unchanged after temperature cycle; (2) 15°; (3) 20°; and (4) 28°. (b) 0.01 μ phosphate-acetate buffer, *pH*=5.9, concentration 0.2 mg/ml, 5 millimeter path length. Solutions containing 25% glycerol show identical behavior. From JAENICKE and LAUFFER (18). Reprinted by permission of the copyright owner, The American Chemical Society

curves and optical rotatory dispersion curves can be distorted by light scatter. Furthermore, JAENICKE (22) has shown, as illustrated in Fig. 4, that the change in optical rotatory dispersion when tobacco mosaic virus protein polymerizes can be duplicated by adding aluminum oxide to a solution of unpolymerized protein, thereby increasing the turbidity. When the aluminum oxide is removed the dispersion of unpolymerized protein is restored. Since an increase in turbidity accompanies polymerization, it is possible that the observed change in optical rotatory dispersion is an optical artefact. Subsequent circular dichroism studies have been reported (23, 24). A shift in mean residue ellipticity at 280 mμ is correlated with polymerization of TMV protein near *pH* 7 where the turbidity effect is negligible (24).

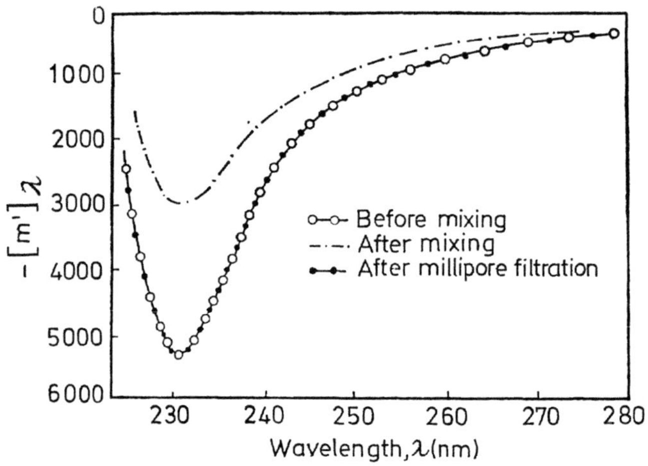

Fig. 4. The effect of turbidity on the optical rotatory dispersion of TMVP. o, normal spectrum at a concentration of 1.8 mg/ml in 0.1 μ phosphate buffer, *pH*=7.0, 27°C. After mixing the solution in the cuvette with Al_2O_3, the depth of the trough is diminished (-.-.-). Filtration leads back to the initial spectrum (•). From JAENICKE (22). Reprinted by permission of the copyright owner, The Federation of European Biochemical Societies

Summary

Many important biochemical reactions, such as the polymerization of tobacco mosaic virus protein, and the formation of gel structures essential for biological processes like cell division are synthetic reactions involving an increase in partial molal volume. From thermodynamics, it is clear that such processes can be spontaneous, i.e., can have a negative free energy change, only when there is a positive entropy change large enough so that, when multiplied by absolute temperature, it exceeds in numerical value the positive enthalpy change. Such reactions are, therefore, entropy-driven. Because of the increase in partial molal volume, thermodynamic deduction leads to the conclusion that these entropy-driven structure-forming processes can be reversed by the application of hydrostatic pressure.

Most of the biologically interesting entropy-driven structure-forming processes involve the polymerization of protein molecules. The polymer molecules interact with each other in some cases to form gels. The explanation common to all of these processes is that side chains of various sorts found in proteins interact with water. A number of different kinds of side chain which interact with water have one feature in common; they all result in a decrease in volume, presumably related to closer packing of water molecules in the region of interaction than in bulk water. These interactions might involve either a positive or a negative free energy, depending upon the specific nature of the side chains and upon the temperature. The free energy of interaction with water of an element on the surface of a protein molecule will depend on the net free energy of interaction of the various side chains that occur on that element.

If, at a particular temperature, that composite free energy of interaction with water is positive, there will be a decrease in free energy if that element of surface can be removed from water. Polymerization of protein molecules or cross-linking of polymerized protein molecules to form gels represent ways in which elements of surface can be removed from water. In general, the entropies of interaction of protein side chains with water are negative and enthalpies negative, just as in the case of the formation of ice from water. If the negative entropy of interaction times the absolute temperature at a particular temperature is not sufficiently high to give a positive free energy for interaction, the contribution of this term can be increased by increasing the temperature. Polymerizations or other reactions brought about as a result of unfavorable interaction of a surface with water are frequently described in terms of hydrophobic bonds; a more accurate name is entropic unions.

Chapter 2

The Role of Water

Introduction

As stated in the first chapter, for many years it was assumed that the increase in entropy needed for entropy-driven structure-forming processes comes from release of bound solvent molecules. Experimental evidence confirming this theory was first obtained by STEVENS and LAUFFER (1) and, more recently, by JAENICKE and LAUFFER (2) in studies on the polymerization of tobacco mosaic virus protein, TMVP. Since the method used to demonstrate water release was thermodynamically based, involving membrane equilibria, it is necessary to have a thermodynamic definition of water-binding or hydration.

Thermodynamic Definition of Hydration

The simplest solution for which one can make thermodynamic measurements of hydration is a ternary solution composed of a high molecular weight, "colloidal" component with zero net charge and two uncharged low molecular weight components. An example would be water, isoelectric protein and glycerol. For this system, hydration is defined thermodynamically as the amount of water which must be added together with unit amount of anhydrous protein to an aqueous glycerol solution so that the change in the logarithms of the activity coefficients of "free" (original) water and glycerol will be equal (3). Essential to the definition is the convention that the *added* water and added anhydrous protein be considered a single component, hydrated protein. The rationale for this definition and its generalization for more complex systems follow.

A binary solution such as n_3 moles of glycerol, component 3, dissolved in n_1 moles of water, component 1, can be characterized by any two of the following:

the mole fractions, x_1 and x_3, the activity coefficients, f_1 and f_3 and the activities, a_1 and a_3. On the rational basis, $a_i \equiv f_i x_i$; $x_i \equiv n_i/(n_1 + n_3)$. If one now adds n_2 moles of anhydrous isoelectric protein, component 2, to this solution, the mole fraction of components 1 and 3, x_1 and x_3, will be reduced by equal fractions; now $x_i \equiv n_i/(n_1 + n_2 + n_3)$. The result will be that the ratio of the mole fraction of component 1 after addition to that before, x_1'/x_1'', will be exactly the same as the comparable ratio for component 3, x_3'/x_3''; $x_i'/x_i'' = (n_1 + n_3)/(n_1 + n_2 + n_3)$. This will be true regardless of the nature of the interaction between the added component 2 and the other components.

In the usual case, the molecules of newly added component 2 will interact differently with the molecules of components 1 and 3 already present. For example, the molecules of component 2 might actually combine chemically with some of the molecules of component 1, or they might attract molecules of component 1 more than those of component 3, or they might repel component 3 molecules more than component 1 molecules. These and many other possibilities are imaginable. What will be observed externally is that the activities of components 1 and 3 will be changed differently, that is, by different fractions, as a result of the addition of anhydrous component 2. It will be recalled that activity is a measure of escaping tendency as exemplified by vapor pressure. Mathematically, this result is expressed by the non-equality, $a_1'/a_1'' \neq a_3'/a_3''$.

The unequal activity ratios can be made equal by the addition of Δn_1 moles of component 1. Conceptually, at least, this quantity can be positive or negative. Added amounts of component 1 will increase a_1 and decrease a_3. Therefore, a value of Δn_1 can be found which makes a_1'/a_1'' equal a_3'/a_3''. The value of Δn_1 moles of component 1 which must be added along with n_2 moles of anhydrous component 2 to produce this result constitutes one form of the definition of the "hydration" of component 2 (3). One way of expressing hydration is in terms of moles of water per mole of anhydrous component 2, $\Delta n_1/n_2$. This definition does not imply that $\Delta n_1/n_2$ is independent of x_2; it applies strictly to a specific solution of definite composition. Another way, more convenient for our purposes, is to express hydration as grams of water per gram of anhydrous component 2, G. Thus, hydration has been defined as the amount of water which must be added along with anhydrous component 2 to result in an equal fractional change in the activities of components 1 and 3.

Chemical potential, μ, is related to activity, a, by the equation,

$$\mu = \mu_x^{\circ} + RT \ln a \tag{1}$$

where R is the gas constant, T is the absolute temperature and μ_x° is a constant independent of composition but dependent upon temperature and pressure. Therefore, at constant temperature and pressure

$$\Delta\mu = RT\, \Delta \ln a \tag{2}$$

The equality of activity ratios for components 1 and 3 can be written,

$$\Delta\ln a_1 = \Delta\ln a_3 \tag{3}$$

Therefore, it is equivalent to $\Delta\mu_1$ equals $\Delta\mu_3$. Thus, for the simple system we are considering an alternative statement of the definition of hydration is that amount of water which must be added along with anhydrous component 2 to result in equal changes in the chemical potential of components 1 and 3.

The activity of the components of a solution is an experimentally determinable parameter. However, the mole fraction is a quantity of which the value is dependent upon how components are defined. The mole fractions of components 1 and 3 before the addition of anhydrous component 2 and Δn_1 moles of component 1 are, $x_1'' \equiv n_1/(n_1 + n_3)$ and $x_3'' \equiv n_3/(n_1 + n_3)$, respectively. After the addition of anhydrous component 2 and the additional component 1, these two mole fractions *can be described* as $x_1' \equiv (n_1 + \Delta n_1)/(n_1 + \Delta n_1 + n_2 + n_3)$ and $x_3' \equiv n_3/(n_1 + \Delta n_1 + n_2 + n_3)$, respectively. If the mole fractions are described in this way, then it is obvious that x_1'/x_1'' will not equal x_3'/x_3''. However, *an equally accurate description* of the solution after the addition of the n_2 moles of anhydrous component 2 plus the Δn_1 moles component 1 is that it contains n_1 moles of "free" component 1, n_2 moles of hydrated component 2 (hydrated to the extent of $\Delta n_1/n_2$ moles per mole) and n_3 moles of component 3. *By this mode of description*, in the final solution x_1' is $n_1/(n_1 + n_2 + n_3)$ and x_3' is $n_3/(n_1 + n_2 + n_3)$. By simple algebraic manipulation, it follows that x_1'/x_1'' equals x_3'/x_3''. Since a equals $f\,x$, it follows that f_1'/f_1'' equals f_3'/f_3'' or $\Delta\ln f_1/\Delta n_2$ equals $\Delta\ln f_3/\Delta n_2$. Thus, a second alternative way of stating the definition of hydration is that it is the amount of component 1 which must be added along with anhydrous component 2 to satisfy the relationship that there be equal changes in the logarithms of the *activity coefficients* of components 1 and 3, both calculated on the basis of considering the added component 1 and the added anhydrous component 2 as a single new hydrated component 2.

The last mode of expressing the definition of hydration is more convenient than either of the previous mathematically identical expressions when one considers the problem of a charged colloidal particle in a system containing dissociated salts. The reasons are inherent in the theory of Donnan equilibrium. There is a further advantage to this alternative way of looking at the problem. According to an argument which will not be reproduced here, it was shown that for an osmotically ideal protein solution, f_1'/f_1'' and f_3'/f_3'' are both precisely equal to one, and even for a protein solution which departs markedly from osmotic ideality, these ratios are so close to one that the deviations are less than the experimental error involved in measuring activity coefficients (3). This important fact makes it possible to state that the activity coefficients of "free" component 1 and of component 3 when their mole fractions are calculated on the "free-water" basis are the same in a solution containing hydrated protein as they are in a simple binary solution containing no protein but the same amounts of components 1 and 3. Physico-chemical data are available from which such activity coefficients can be calculated.

It should be observed that, in the example presently being considered, it would be possible to cause a_1'/a_1'' to equal a_3'/a_3'' by adding both component 1 and component 3. An infinite number of pairs of values of Δn_1 and Δn_3 would bring about this equality and therefore establish the condition that $\Delta \ln f_1$ equals $\Delta \ln f_3$. The value chosen as the definition of hydration was the value of Δn_1 corresponding to Δn_3 equal to zero. When a system with more than three components is considered, one must satisfy the condition that $\Delta \ln f_1 = \Delta \ln f_3 = \Delta \ln f_4$, etc. This condition can be brought about by adding components 1, 3, 4 etc. to the system, and the hydration can be considered to be the value of Δn_1, which corresponds to the case in which one of the values, Δn_3, Δn_4, etc., is zero and the others positive. To state this less abstractly, in a system containing water, protein, glycerol and electrolyte, one can determine a value for water-binding only by simultaneously establishing a value for either glycerol binding or electrolyte binding.

The Measurement of Changes in Hydration

STEVENS and LAUFFER (1) measured the water released, that is, the change in hydration, during the polymerization of TMVP by using a quartz spring-balance, illustrated diagramatically in Fig. 1. The apparatus consists of a cellophane dialysis sac attached to a delicate quartz helix. Inside the sac, one places protein dissolved in water, glycerol and electrolyte (buffer). The sac is suspended in a water-glycerol-electrolyte solution. Roughly, the idea inherent in this method is that "free" water, glycerol, and electrolyte are free to diffuse through the membrane so that the weight of each is cancelled by buoyancy. Protein and "bound" water, however, are not able to pass the membrane and, as

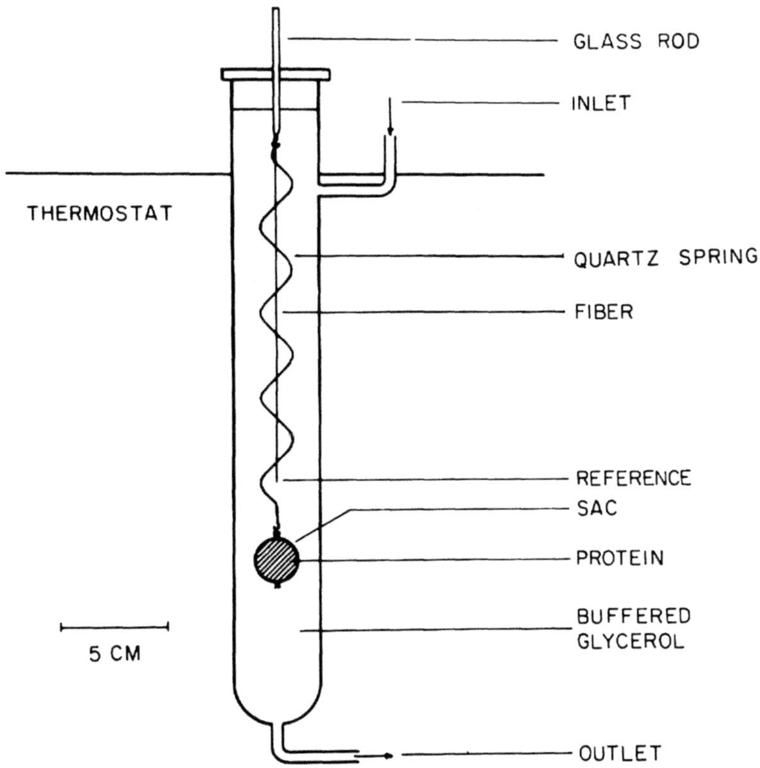

Fig. 1. Schematic diagram of spring-balance showing dialysis sac, quartz spring and chamber. From STEVENS and LAUFFER (1). Reprinted by permission of the copyright owner, The American Chemical Society

a result, will be weighed, corrected for buoyancy. If "bound" water is released upon polymerization, it will come to equilibrium and will no longer be weighed. The reason for having glycerol in the system is to increase the density of the solution, for, otherwise, the weight of "bound" water would be cancelled by buoyancy.

In reality, the theory of the method is more complex than that just described. It is necessary to understand precisely the membrane equilibria involved before valid measurements of change in hydration can be made. Because the cellophane sac contains a non-dialyzable component, the protein, equilibrium is established in which the contents are at a higher pressure than the external solution, the difference being the colloidal osmotic pressure. If the protein is charged, Donnan equilibrium will result in different electrolyte concentrations inside and outside the sac, a factor which would contribute to the weight. There is the additional possibility of electrolyte binding by the protein. The actual binding of electrolyte and the redistribution of electrolyte by Donnan equilibrium can be linked together as a single term. Finally, there is the possibility that the glycerol concentrations will not be the same on the two sides of the membrane at equilibrium. To decide whether or not a significant contribution to the weight, or at least to the change in weight associated with polymerization, arises from this source, it is necessary to consider the problem of membrane equilibria in multi-component systems. The reader willing to assume that this is negligible can omit the next paragraph.

In a multi-component membrane system at equilibrium, every component except the non-dialyzable component 2 must satisfy the Gibbs condition, expressed in the integrated form by:

$$\ln a''_i/a'_i = \pi \, \overline{V}_i/RT = \phi \, m_2 \, \overline{V}_i \tag{4}$$

where π is the osmotic pressure, ϕ is the osmotic coefficient, $\overline{V}_i$ is the appropriate average partial molar volume of the ith constituent and m_2 is the molarity of component 2. When this equation is applied to components 3 and 1, the result is;

$$(a_3'/a_1') = (a_3''/a_1'')\ e^{-\phi \, m_2(\overline{V}_3 - \overline{V}_1)} \simeq (a_3''/a_1'')(1 - \phi m_2(\overline{V}_3 - \overline{V}_1)) \tag{5}$$

The approximation is accurate because the exponent is very small. Because $a_3/a_1 = (x_3 f_3/x_1 f_1) = (m_3 f_3)/(55.6 f_1)$, one obtains,

$$m_3' = m_3'' (f_1' f_3''/f_3' f_1'')(1 - \phi m_2 (\overline{V}_3 - \overline{V}_1)) \tag{6}$$

The solution of the membrane equilibrium problem was found in this form by ADAIR in 1937 (4). However, at that time the meaning of the activity coefficients of water and glycerol in a solution containing hydrated protein was not understood. The advantage of the thermodynamic definition of hydration presented above is that it makes it possible to assign meanings to these activity coefficients, and thereby to calculate the glycerol concentration in the protein, m_3', from the known concentration in the solution outside the membrane, m_3''. STEVENS and LAUFFER (1) and JAENICKE and LAUFFER (2) have made such calculations and have found that, under the experimental conditions used in their studies, the difference between m_3' and m_3'' is so small that this factor can be neglected entirely without introducing serious error. For the sake of simplicity, therefore, the redistribution of glycerol will not be taken into account in the subsequent development of the theory for the spring-balance experiments. However, if, in other experiments, this term should prove not to be negligible, it can be dealt with by computation.

Taking into consideration the two simplifications descriped above, namely, that redistribution of electrolyte in accordance with Donnan equilibrium will be considered as a contribution to salt-binding and that redistribution of glycerol as a result of the membrane equilibrium contributes negligibly, Eq. (7) can be written for the weight corrected for buoyancy, W, of a solution in the sac containing γ_2 grams of anhydrous protein. The sac will also contain water, glycerol and electrolyte in equilibrium with the external solution.

$$W = \gamma_2 G (1 - V_1 \rho) + \gamma_2 (1 - V_2 \rho) + \gamma_2 G' (1 - V_4 \rho) \tag{7}$$

In Eq. 7, G means the hydration of protein expressed as grams of water per gram of anhydrous protein and G' means "bound" salt expressed as grams per gram. The symbol, ρ, refers to the density of the external solution and V_1 V_2 and V_4 refer to the partial specific volumes of water, protein and electrolyte, respectively. The first term on the right is the buoyant weight of the "bound" water. The third term is the buoyant weight of the "bound" salt. If the state of polymerization of the protein (TMVP in the experiments considered) is changed, for example by changing pH, without changing the density of the solution outside the sac, conceptually three additional changes can occur with respect to the protein: change in water bound, change in salt bound, and change in partial specific volume. Thus, one can write for the change in weight, ΔW, accompanying polymerization,

$$\Delta W/\gamma_2 = \Delta G(1 - V_1\rho) - \Delta V_2\rho + \Delta G' (1 - V_4\rho) \quad (8)$$

There is in fact a change in partial specific volume of the protein when it polymerizes. The change in partial specific volume, ΔV_2, can be measured in an independent experiment involving a Linderström-Lang dilatometer, like that illustrated diagramatically in Fig. 2. If unpolymerized protein is placed in one arm of the inverted V-shaped tube and some chemical which will cause the protein to polymerize, such as, for example, an acid, is placed in the other arm and then the balance of the apparatus is carefully filled with solvent into the capillary, the change in partial specific volume can be determined by measuring the change in the volume of the solution as indicated by the change in the level in the capillary when the chemical and the protein are mixed to bring about polymerization. This volume change divided by the mass of protein in the vessel is the change in partial specific volume.

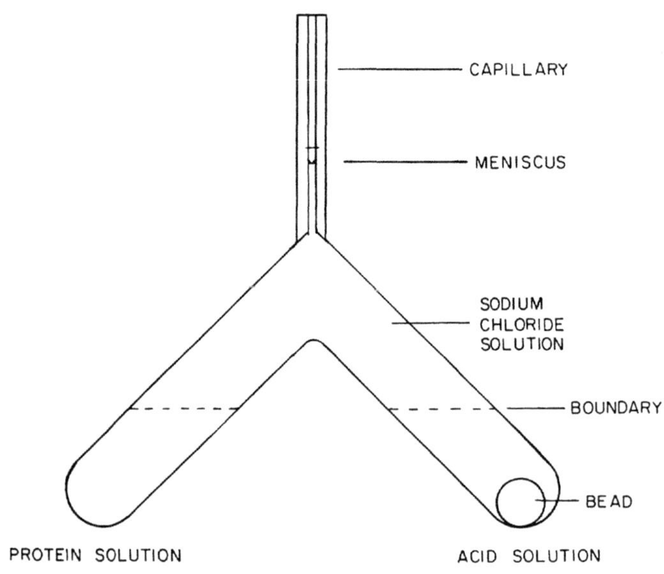

Fig. 2. Dilatometer for determining volume change accompanying polymerization of TMVP. From LAUFFER (5). Reprinted by permission of the copyright owner, The Avi Publishing Co., Inc.

With ΔV_2 evaluated, both ΔG and $\Delta G'$ can be evaluated by measuring $\Delta W/\gamma_2$ in media of different densities. When the density of the medium, ρ, is $1/V_1$,

$$\Delta W/\gamma_2 + \Delta V_2\rho = \Delta G' \; (1 - V_4\rho) \tag{9}$$

This equation can be solved for $\Delta G'$, since all other quantities can be measured. On the assumption that $\Delta G'$ does not change when the density of the medium is changed by adding glycerol, ΔG can be evaluated by using Eq. (8) in a second experiment in which the density is increased by adding glycerol.

Volume Change in TMVP Polymerization

Tobacco mosaic virus, TMV, is composed of 95% protein and 5% RNA. The RNA in a TMV particle is a single thread with a molecular weight of about 2 million. The protein component is composed of identical subunits with a molecular weight of about 17,500. In the virus, the protein subunits and the RNA thread are held together in right-handed (6) helices, as illustrated in Fig. 3. When the protein is isolated from the virus and is dissolved in buffers with *pH* values between 7 and 8, it is dispersed in a slightly polymerized state as trimers, hexamers, etc. of the subunit at all temperatures between 0 and 25°C. The exact state of dispersion depends on concentration, temperature, *pH* and ionic strength, μ. When the *pH* is reduced to 5.5, the protein exists in the temperature range mentioned above in a highly polymerized state, in which the protein subunits are arranged in a helix like that found in the virus. The main difference between the structure of polymerized protein at *pH*=5.5 and that of virus is the absence of the RNA helix in the former. At *pH*=6.5, 0.1 μ, the protein is in the slightly polymerized state at 4°C and in the highly polymerized state at 20°C. There are other modes of polymerization of TMV protein which will be discussed in subsequent chapters. However, for the present purpose it is sufficient to know that polymerization from a slightly polymerized state into high molecular weight rods with a helical structure resembling that found in the virus can be brought about by changing *pH* of 0.1 μ buffers from 7.5 to 5.5 at 4°C or 20°C, and that in a buffer of the same ionic strength at *pH*=6.5, the polymerization can be brought about by changing the temperature from 4°C to approximately 20°C.

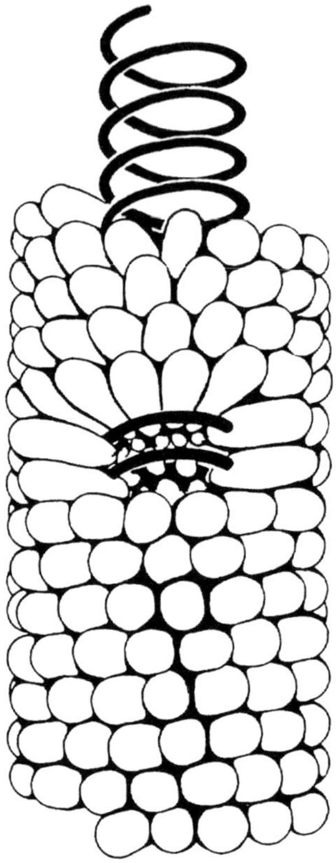

Fig. 3. The structure of tobacco mosaic virus. Inner helix represents RNA, and helically assembled ellipsoids of revolution represent protein subunits. Drawing based on a model originally constructed by Dr.R. FRANKLIN. The helix - is represented as right-handed, conforming to the evidence of FINCH (6).

The change of partial specific volume upon polymerization of tobacco mosaic virus protein, TMVP, was measured by STEVEN and LAUFFER (1) in a Linderström-Lang dilatometer, as illustrated in Fig. 2, with a volume of 32 ml. Seven ml of protein solution in 0.1 m sodium chloride adjusted to *pH*=7.6 was introduced into one leg of the dilatometer. In the other leg was placed a solution 0.1 m with respect to sodium chloride, with sufficient hydrochloric acid to titrate the protein to *pH*=5.5 when mixed. The liquid in the two legs of the dilatometer was then layered with a 0.1 m solution of sodium chloride in an amount sufficient to fill the vessel into the capillary. After thermal equilibration in a 4.0°C bath, the position of the meniscus in the capillary was determined with a cathetometer. The liquids were then mixed, and the position of the meniscus was remeasured at 4°C. The change in volume observed in the previously calibrated capillary tube divided by the mass of protein is the change in partial specific volume of the protein brought about by this operation. Control experiments were carried out with TMV, which does not change its state of polymerization drastically over the same *pH* range. A further control involved diluting protein in the dilatometer without acid. The results as presented originally by STEVENS and LAUFFER (1) are shown in Table 1.

Table 1. Change of apparent partial specific volume of virus and protein[a]

	Virus	Protein	Protein (dilution)
Δh (mm)	+ 0.27	+ 1.21	- 0.061
	+ 0.32	+ 1.23	
w (g)	0.169	0.128	0.105
	0.169	0.131	
ΔV (cc/g)	+ 0.00127	+ 0.00744	- 0.00046
	+ 0.00149	+ 0.00738	
Final *pH*	5.3	5.5	
	5.3	5.5	
$\Delta V_a - \Delta V_b$	+ 0.0060 (cc/g)		

[a] A double entry represents an experiment in duplicate. The diameter of the capillary bore of the dilatometer is 1.0 mm; Δh is the change of height of the meniscus (+ represents an expansion), w is the weight of material (virus protein), and ΔV is the change of apparent partial specific volume. From STEVENS and LAUFFER (1). Reprinted by permission of the copyright owner, The American Chemical Society.

The experiment shows that there is an average increase in partial specific volume of the protein of 0.00741 ml/g, when polymerization is brought about in this manner. The experiment, however, unavoidably involved simultaneous dilution of 0.13 g of protein in 7 ml to 0.13 g in 32 ml. The control experiment shows that upon dilution alone, V_2 decreases 0.00046 ml/g. This decrease is probably attributable to a great extent to a decrease in the state of polymerization of the slightly polymerized protein at *pH*=7.6. Thus, if it had been possible to bring about polymerization by changing *pH* from 7.6 to 5.5 without changing protein concentration, that is, by keeping it constant at 0.13 g in 32 ml or 0.004 g/ml, the increase in V_2 would have been 0.00787 ml/g. A slightly higher value should be obtained for polymerization at even lower constant protein concentrations.

Acid-base titration studies carried out by SCHEELE and LAUFFER (7) show that TMV protein, TMVP, binds approximately 3 hydrogen ions per protein subunit of molecular weight 17,500 when the *pH* is reduced from 7.6 to 5.5. Similar experiments showed that TMV binds almost the same amount of hydrogen ion when titrated over the same *pH* range. Because of electrostriction, hydrogen ion binding should contribute to the increase in volume. The extent can be estimated from the control experiment involving TMV. In this experiment, V_2 increased on the average 0.00138 ml/g. This control experiment also involves dilution along with change in *pH* and hydrogen ion binding. JAENICKE and LAUFFER (2) estimated that the change in volume for diluting TMV is approximately +0.00009 ml/g. Thus, the volume change with TMV, associated with hydrogen ion binding only, is approximately +0.0013 ml/g. This should be a very good estimate of the volume change attributable to electrostriction when protein solution is changed from *pH*=7.6 to 5.5. Therefore, there must be an increment in V_2 of 0.0066 ml/g attributable to something other than hydrogen ion binding when TMVP is caused to polymerize in this manner.

Additional experiments on TMV and TMVP were performed in the same manner with the same dilatometer by JAENICKE and LAUFFER (2), except that the temperature was 25°C and the electrolyte was 0.1μ KCl. The results are shown in Table II. The most important result is that when polymerized protein at 25°C is changed from *pH* values between 6.7 and 6.85 to around 5.5, there is an increase in V_2 of 0.00247 ml/g, on the average.

Table 2. Dilatometric effect of Charge on TMV and Polymerized TMV-P, $T = 25^{o}$, $\mu = 0.1$ (KCl)

	Material (mg)	pH Inital	pH Final	Δh (mm)	$\Delta V'$ (cc/g)	$\Delta V^* = \Delta V' - \Delta V_{dil}'$ (cc/g)
TMVP	87	6.70	5.45	0.309	0.0028	0.0026
	96	6.70	5.40	0.291	0.0024	0.0022
	385	6.77	5.30	1.347	0.0028	0.0026
	300	6.85	5.62	0.737	0.0019	0.0017
TMVP $\Delta V_{dil}'^{a}$	370	6.66	6.66	0.054	0.0001_9	
TMV	282	6.80	5.52	0.325	0.0009_1	0.0008_2[b]
TMVP-TMV $\Delta V_a' - \Delta V_b'$		6.8	5.5		0.0015_7[c]	

[a] $\Delta V_{dil}'$ means $\Delta V'$ on dilution. [b] $\Delta V_{dil}'$ for TMV is too small to be given precisely: 0.0000_9 cc/g. [c] Changes of *pH* from 7.5 to 5.5 which include both the "charge effect" and polymerization give 0.005_9 cc/g. From JAENICKE and LAUFFER (2). Reprinted by the permission of the Copyright owner, The American Chemical Society.

One of the questions that can be answered from the combined data of Tables 1 and 2 is the approximate magnitude of ΔV_2 when concentrated unpolymerized protein at *pH* 7.6, 0.1 μ at 25°C is polymerized by titrating to *pH* 6.8 without change in concentration. This will be the difference between the average value of ΔV_2 not corrected for dilution found in Table 1, 0.00741 and the average value for ΔV_2 found in Table 2, 0.00247 ml/g. This amounts to 0.0049 ml/g. To make this calculation, it was necessary to assume that the results shown in Table 1 would have been the same had the experiment been carried out at 25°C instead of 4°C. This is a reasonable assumption, because JAENICKE and LAUFFER (2) showed that the temperature coefficient of the partial specific volume of unpolymerized protein was equal, within experimental error, to that of TMV and, therefore, presumably of TMVP over the entire temperature range considered here. SCHEELE and LAUFFER (7) found that unpolymerized protein does not bind hydrogen ions between *pH*=7.6 and 6.8, and also that hydrogen ions are bound during the process of polymerization. Thus, the hydrogen ions bound over this range of *pH* are legitimately part of the polymerization process when polymerization is brought about by increasing the temperature at a constant *pH* of 6.8. Therefore, the figure 0.0049 ml/g is also the best estimate of the change in partial specific volume when polymerization is brought about by increasing temperature at constant *pH*, a quantity needed for interpreting the spring balance experiments of JAENICKE and LAUFFER (2).

Another question that can be answered from the combined data of Tables 1 and 2 is the approximate magnitude of ΔV_2 when concentrated unpolymerized protein at *pH*=7.6, 0.1 μ at 4°C is polymerized by titrating to 5.5 without change in concentration, as in the spring-balance experiment of STEVENS and LAUFFER (1). Here again, the assumption, justified above, that changes in volume are independent of temperature over the range considered must be made. The additional assumption must be made that the volume change associated with diluting polymerized protein is the same at *pH*=5.5 as at *pH*=6.66. The sought value of ΔV_2 can be obtained by subtracting 0.00019 ml/g found in Table 2 for diluting polymerized protein[1] from the average value for protein shown in the third

---

[1] This is probably an over-correction because the protein was more concentrated in the experiment reported in Table 2 than in that reported in Table 1. Because the total correction is only 2.5%, over-correction leads to no large error.

column of Table 1, 0.00741. Thus, the best estimate of the value ΔV_2 for the transition under discussion is 0.00722 ml/g.

The fact that there is an increase in partial specific volume when TMVP polymerizes is of intrinsic interest. From it, one can reason thermodynamically in accordance with Eq. (17) of Chap. 1, that an increase in pressure will shift the equilibrium between polymerized and unpolymerized TMVP toward the side of the less polymerized state. The increase in partial specific volume can be interpreted on the assumption that during the polymerization process, water molecules packed closely around protein surfaces are released to become less closely packed free water.

The increase in partial specific volume of polymerized TMVP when the *pH* is changed from 6.8 to 5.5 is also of considerable interest. Part of this change is attributable to hydrogen ion binding. According to SCHEELE and LAUFFER (7), TMV and TMVP bind approximately the same amount of hydrogen ion between *pH*=6.8 and *pH*=5.5. Therefore, the portion attributable to hydrogen ion binding can be estimated from the volume increase for TMV over this range, 0.009 ml/g. The difference between this and the average value shown in Table 2 for TMVP, 0.0016 ml/g, must stem from something other than hydrogen binding and the accompanying change in electrostriction. The interpretation advanced by JAENICKE and LAUFFER (2) is that as *pH* is reduced with polymerized protein, there is a decrease in net negative charge and therefore lowered electostatic repulsion between protein subunits. This permits tighter packing of the protein subunits in the polymer structure, thereby allowing for additional release of water and a parallel increase in partial specific volume.

Water Release upon Polymerization of TMVP

The first attempt to measure the water released when TMVP polymerizes was the spring-balance experiment of STEVENS and LAUFFER (1). In it, 0.10 g of TMVP dissolved in approximately 0.1 μ sodium chloride solution buffered to *pH*=7.50[2], containing 0.25 g glycerol per ml, was placed inside the sac. The tube outside contained a water-glycerol-salt solution of the same *pH*, ionic strength and glycerol content. The extension of the spring was then noted after thermal equilibrium had been reached at 4.0°C. When the spring-balance had reached a stable extension, the external solution was replaced by one of approximately the same density, but buffered to *pH*=5.5. After sufficient time to permit equilibration of the electrolyte across the membrane, with the resultant drop in *pH* and polymerization in the protein solution, the extension of the spring was again read. Finally, to demonstrate reversibility, the original solution at *pH*=7.5 was placed in the external tube, and sufficient time was allowed for equilibrium to be established. The results are shown in Fig. 4. It can be seen that the spring extension was completely reversible. The change in weight was calculated from the difference between the original spring extension and the stable value achieved after equilibration at *pH*=5.5. The result is recorded in Table 3. Another experiment was carried out with TMVP in salt solution containing no glycerol. In order to be able to subtract water release associated with hydrogen ion binding, identical experiments were performed on TMV, which binds very nearly the same amount of hydrogen ion between *pH*=7.5 and 5.5 as does TMVP. The results of all these measurements are shown in Table 3.

According to the simplified theory outlined in a preceding section, one can write Eq. (8) for each experiment. The weight of protein or virus, γ_2, was the same in all experiments and controls. For the experiment on TMVP and the control on TMV involving the glycerol solution:

$$\Delta\Delta W/\gamma_2 = \Delta\Delta G\ (1 - V_1\rho) - \Delta\Delta V_2\rho + \Delta\Delta G'\ (1 - V_4\rho) \qquad (10)$$

---

[2] The concept of *pH* applied to a glycerol solution presents difficulties; *pH* cannot have the same meaning as it has in water. Strictly speaking, *pH* of a glycerol solution simply means the reading on a *pH* meter, the value of which affects TMVP polymerization.

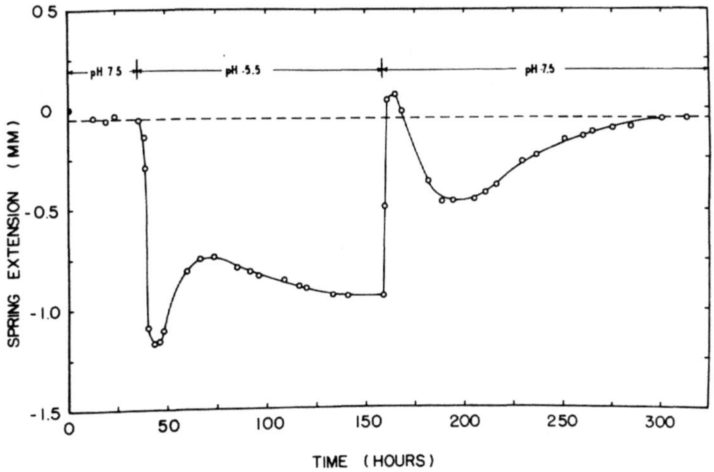

Fig. 4. The time course of equilibration of TMVP in the spring-balance. The protein was equilibrated successively to buffered glycerol solutions with a *pH* of 7.5, 5.5 and again 7.5. From STEVENS and LAUFFER (1). Reprinted by permission of the copyright owner, The American Chemical Society

Table 3. Tabulation of the change of apparent weight of the virus and protein systems in the quartz spring balance [a]

	Virus (mg)	Protein (mg)
Buffer solution (density 1.0044)	-0.338	-0.638
Buffered glycerol solution (density 1.0665)	-0.204	-0.368

[a] Each entry gives the change of weight of the sac between *pH* = 7.5 and 5.5. Each sample contained 0.1 g of virus or protein. From STEVENS and LAUFFER (1). Reprinted by permission of the copyright owner, The American Chemical Society.

The meaning of the symbol $\Delta\Delta$ is the difference between the value of Δ obtained with protein and with control virus. An identical equation can be written for the experiment and control carried out in the solution containing no glycerol. Thus, one has two simultaneous equations in which all quantities except $\Delta\Delta G$ and $\Delta\Delta G'$ can be determined by experiment. It is therefore possible to solve for $\Delta\Delta G$ and $\Delta\Delta G'$.

It was shown in the last section that the best estimate of the value of ΔV_2 for TMVP when pH is changed at constant concentration from approximately 7.6 to 5.5 is 0.00722 ml/g. By subtracting the value for dilution of TMV, 0.00009 ml/g from the average value of ΔV for TMV shown in Table 1, the best estimate of ΔV_2 for titrating a concentrated solution of TMV from pH=7.5 to about 5.5 is obtained, 0.00129. The best estimate of $\Delta\Delta V_2$ is therefore 0.00722 - 0.00129 or 0.00593. The values of ρ for the solutions containing 25% glycerol and no glycerol, respectively, were 1.0665 and 1.0044. The values of V_1 and V_4 are taken to be 1.000 and 0.290, respectively. The value of γ_2 is 0.10 for all experiments and controls. The values of $\Delta\Delta W$ for experiments in the buffer solution and in the buffered glycerol solution are taken from Table 3. They are -0.638 + 0.338 or -0.300 mg or -0.0003 g and -0.164 mg or -0.000164 g, respectively. The solution of the simultaneous equation then yields $\Delta\Delta G$ = -0.029 g per g and $\Delta\Delta G'$ = +0.0040 g per g. These values are very close to the values calculated by STEVENS and LAUFFER through the use of the exact theory, -0.027 g per g and +0.0043 g per g, illustrating the validity of the approximations made in the simplified theory.

JAENICKE and LAUFFER (2) carried out a series of experiments, similar to those of STEVENS and LAUFFER (1), in which polymerization of TMVP was brought about by raising the temperature. The results of a typical experiment, shown in Fig. 5, demonstrate that there is a definite change in the spring extension when the temperature is increased, that the change is completely reversible and that it is much greater for TMVP, which polymerizes, than for TMV, which does not. A complete experiment involves measuring the change in buoyant weight with temperature of TMVP and the same thing for TMV. The difference between these two is $\Delta\Delta W$. Such experiments were repeated with a number of solutions in glycerol-water mixtures of different compositions and therefore different densities, ρ. The results of all such experiments are shown in Table 4.

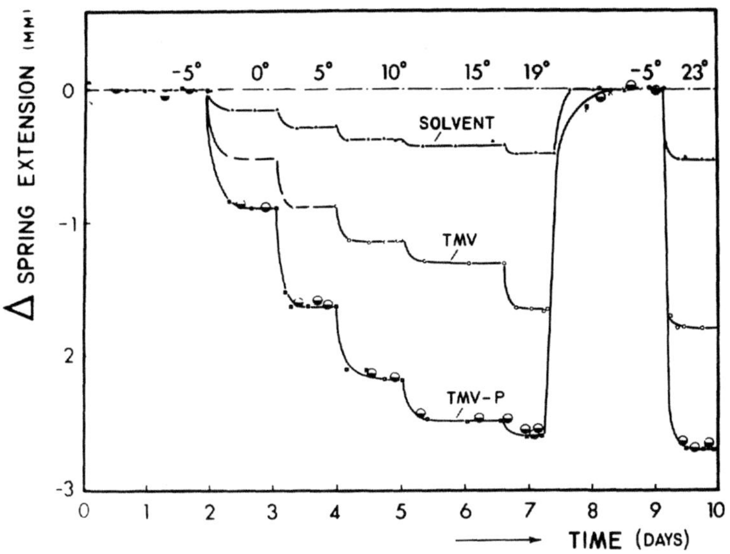

Fig. 5. Time course of equilibration of TMVP in the spring-balance. The total amount of material in the sac; 209 mg TMV (o), 200 mg of TMVP (■), 140 mg of TMVP (O) normalized for 200 mg of material. Solvent is 25 g per 100 ml glycerol in phosphate buffer pH=6.77, μ = 0.05 + 3 x 10^{-3} M EDTA. From JAENICKE and LAUFFER (2). Reprinted by permission of the copyright owner, The American Chemical Society

These experiments are interpreted by using Eq. (7). If subscript a refers to TMVP and b to TMV, T_2 to a high temperature (such as 20°C), and T1 to a lower temperature, the following equation can be written for ΔW_a and a comparable one for ΔW_b.

$$\Delta W_a/\gamma_2 = G_{aT2}\,(1 - V_1\rho)_{T2} - G_{aT1}\,(1 - V_1\rho)_{T1} - (V_{2a}\rho)_{T2} + (V_{2a}\rho)_{T1} + G'_{aT2}\,(1 - V_4\rho)_{T2} - G'_{aT1}\,(1 - V_4\rho)_{T1} \quad (11)$$

$\Delta\Delta W$ is defined as $\Delta W_a - \Delta W_b$.

$$\Delta\Delta W/\gamma_2 = (G_a - G_b)_{T2}\,(1 - V_1\rho)_{T2} - (G_a - G_b)_{T1}\,(1 - V_1\rho)_{T1} - [(V_{2a} - V_{2b})\rho]_{T2} + [(V_{2a} - V_{2b})\rho]_{T1} + (G'_a - G'_b)_{T2}\,(1 - V_4\rho)_{T2} - (G'_a - G'_b)_{T1}\,(1 - V_4\rho)_{T1} \quad (12)$$

Table 4. Changes with temperature in the buoyant weight of TMV and TMV-P in the quartz spring balance. PO_4 - Buffer + 1.5-3.0 x 10^{-3} M Edta

% (w/v) Glycerol	pH	μ	Temp.-Change (°C)	$-\Delta W_{\Delta T}$ TMV	$-\Delta W_{\Delta T}$ TMV-P	$-[\Delta W_{\Delta T} - (\Delta W_{\Delta T})_0]^a$ TMV	TMV-P	$-\Delta\Delta W$
0	6.58	0.05	20→4			0.378	0.679	0.301
			2→20			0.459	0.714	0.255
			25→2			0.458	0.816	0.358
						0.458	1.033	0.575
			20→2			0.440	0.912	0.472
	6.77	0.05	0→18	0.829	1.232			0.403
			0→24	0.931	1.384			0.453
			0→24	0.938	1.386			0.448
10	6.77	0.05	-2→22	0.704	1.030			0.331
			22→-2	0.721	1.039			0.318
18	6.77	0.05	-2→22	0.468	0.716			0.248
			22→-2	0.472	0.711			0.239
25	6.77	0.05	-5→24			0.442	0.663	0.221
			-5→19			0.298	0.531	0.233
			19→-5			0.302	0.505	0.203
			-5→18			0.424	0.571	0.147
	6.50	0.05	-5→22			0.253	0.458	0.205
						0.253	0.492	0.239

[a] $(\Delta W_{\Delta T})_0$ is the change in buoyant weight with temperature when the sac is filled with pure solvent. This can be omitted when the sacs used for TMV and TMV-P are of identical weight. From JAENICKE and LAUFFER (2). Reprinted by permission of the copyright owner, The American Chemical Society.

By algebraic manipulation and regrouping of terms (8), this becomes:

$$\Delta\Delta W/\gamma_2 = (1 - V_1\rho)_{T1}\,\Delta\Delta G - (G_a - G_b)_{T2}\,\Delta V_1\rho - \Delta\Delta V_2\rho_{T2}$$
$$- (V_{2a} - V_{2b})_{T1}\,\Delta\rho + (1 - V_4\rho)_{T1}\,\Delta\Delta G'$$
$$- (G'_a - G'_b)_{T2}\,\Delta(V_4\rho) \qquad (13)$$

The fourth term on the right is negligible because both $\Delta\rho$ and $(V_{2a} - V_{2b})_{T1}$ are very small and the final term on the right should be nearly zero because the sum of true salt binding and Donnan "binding" should be nearly the same for virus and polymerized protein in view of the similarities in hydrogen ion titration curves of the two. Thus, the working equation simplifies to;

$$\Delta\Delta W/\gamma_2 = (1 - V_1\rho)_{T1}\,\Delta\Delta G - (G_a - G_b)_{T2}\,\Delta V_1\rho - \Delta\Delta V_2\rho_{T2} + (1 - V_4\rho)_{T1}\,\Delta\Delta G' \tag{14}$$

As shown in the preceding section, the appropriate value of $\Delta\Delta V_2$ for this experiment is 0.0049 ml/g. The value of V_1, V_4 and ρ for the various solutions at the temperatures $T1$ and $T2$ are available in the literature, or can be determined by independent experiments. The value of $(G_a - G_b)_{T2}$, that is, the difference between the hydration of polymerized TMVP and TMV was determined by independent measurements made with the spring-balance by JAENICKE and LAUFFER (9). The values obtained were 0.20 and 0.16 g water per g protein, respectively, for G_a and G_b. The experimental results from which these hydrations were calculated are shown in Fig. 6. These data can be interpreted

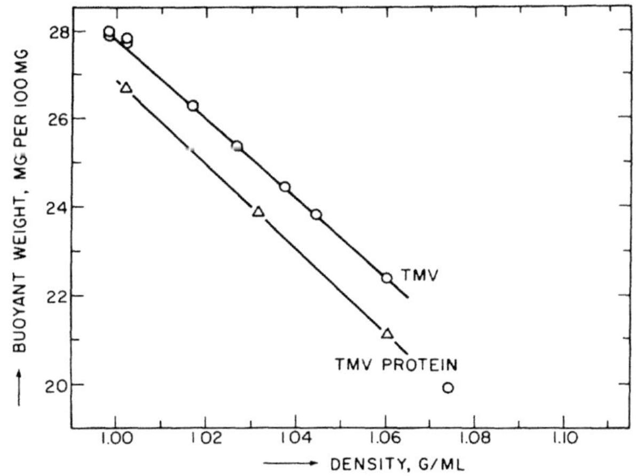

Fig. 6. Buoyant weight W mg per 100 mg as a function of density from spring-balance experiments in aqueous buffers and glycerol buffer mixtures, 20°C. TMV in phosphate buffer, pH=6.77, μ = 0.05. TMVP in phosphate buffer, pH=6.7, μ = 0.05. [Data of JAENICKE and LAUFFER (9)]. From LAUFFER (8). Reprinted by the permission of the copyright owner, Marcel Dekker, Inc.

directly through the use of Eq. (7). If it is assumed that the contribution of ion binding is negligible, that is, that G' is zero, the basic equation can be written:

$$W/\gamma_2 = G(1 - V_1\rho) + (1 - V_2\rho) \qquad (15)$$

One can evaluate V_2 from the value of W for a system in which $\rho = 1/V_1$. In this manner, V_2 for TMV and TMVP were found to be 0.722 and 0.732, respectively. These values are in excellent agreement with values obtained by pycnomometry, 0.727 and 0.734, respectively. This good agreement is evidence that neglecting possible salt binding does not cause appreciable error. After V_2 is thus evaluated, according to Eq. (15), one can evaluate G from the slope of the line obtained when W/γ_2 is plotted against ρ. The slope is equal to $-G\ V_1 - V_2$.

When the solvent contains no glycerol, $1/V_1$ is practically equal to ρ, and $\Delta V_1\rho$ is practically zero. Under these circumstances, Eq. (14) reduces to

$$\Delta\Delta W/\gamma_2 = -\Delta\Delta V_2\rho_{T2} + (1 - V_4\rho)_{T1}\Delta\Delta G' \qquad (16)$$

When the data shown in Table 4 are interpreted in this manner, a mean value of +0.0010 g per g is obtained for $\Delta\Delta G'$. This would correspond to a fraction of a mole of salt per mole of protein monomer. The value is not significantly different from zero. When the data of Table 4 obtained at other values of ρ are then interpreted with Eq. (14), one obtains $\Delta\Delta G = -0.033$ g of water per g of protein (2), in good agreement with the value obtained using the data of STEVENS and LAUFFER (1), -0.029 g per g. In reality, there is no good reason to expect close agreement between the two figures, because the initial and final states are not identical in the two experiments. However, it is significant that the two experiments agree in demonstrating beyond reasonable doubt that water actually is released when TMVP polymerizes.

The spring-balance experiments both of STEVENS and LAUFFER (1) and JAENICKE and LAUFFER (2) indicate that the change in salt binding upon polymerization is small, possibly not significantly different from zero. This conclusion is borne out by ion binding studies of SHALABY *et al.* (10) and of BANERJEE and LAUFFER (11). SHALABY *et al.* measured ion binding by TMV and TMVP with anion and cation-exchange membrane electrodes. At 22°C, TMV bound small amounts of K^+ and Cl^- and larger amounts of Ca^{++} at *pH* values above 7 but none of these at *pH*=4.6. At the same temperature, TMVP does not bind Cl^-, K^+, or Ca^{++}, either at *pH*=4.6, where the protein is polymerized, or at *pH* values between 7.5 and 9, where the protein is unpolymerized. Thus, one would expect no change in the binding of any of these ions when polymerization is brought about by changing *pH* at room temperature. BANERJEE and LAUFFER (11) measured Cl^- and K^+ binding by both TMV and TMVP at *pH*=6.5, at low temperature where the protein is unpolymerized and also at room temperature where it is polymerized. In these experiments, specific ion electrodes were used. Neither ion was bound by TMVP or TMV at either temperature. Thus, all studies made thus far confirm the conclusion that ion binding under the conditions in which TMV protein is polymerized is negligible for both the unpolymerized and the polymerized state and that therefore the change in ion binding is negligible.

Because interaction with water is a very important aspect of the polymerization reaction for TMVP, it is to be expected that substances which alter the structure of water should affect polymerization of this material. In the view of FRANK and EVANS (12), water is represented by an equilibrium between simple molecules and aggregates of a limited degree of order. Some substances are reputed to break the structure of water. Among these are thiourea, urea, dioxane and acetamide. In studies carried out by SMITH and LAUFFER (13) and by SHALABY and LAUFFER (14), all of these uncharged structure-breakers were found to favor depolymerization of TMVP. Sucrose is reputed to be a structure-maker; it was found in these studies to favor polymerization of TMVP. D_2O is regarded as having greater structure than H_2O. The substitution of D_2O for H_2O fits the same pattern; KHALIL and LAUFFER (15) found that TMVP polymerizes more readily in D_2O than in H_2O.

The release of water molecules in the polymerization of TMVP demonstrated here replaces theory with a validated reason to explain the increase in entropy needed to drive this particular entropy-driven structure-forming process. It seems valid to generalize and conclude that other endothermic, entropy-driven structure-forming biological processes derive their increase in entropy from release of water molecules, thereby focusing attention on the importance of water in biological structures. Water plays a formerly unappreciated role in living systems. R.A. GORTNER (16) seems to have anticipated something of our current understanding when he wrote in 1930; "It is my belief that many of the reactions characteristic of living processes have to do more with the water relationships of the organism than with any other single factor."

Summary

For many years it has been the theory that the increase in entropy which drives endothermic structure-forming processes in biological and biochemical systems stems from the release of solvent molecules during polymerization. This theory has been demonstrated to be correct in the case of the polymerization of tobacco mosaic virus protein.

Since the method of demonstrating water-release is a thermodynamic one, it is necessary to have a thermodynamic definition of hydration. Thermodynamically, hydration is defined as that amount of water which must be added, along with a given amount of anhydrous protein, to a binary solution, such as water and glycerol, so that the logarithms of the activities or of the activity coefficients of water and low molecular weight solute are changed equally. One of the consequences of this definition is that the activity coefficients of free solvent and low molecular weight solute are very nearly identical in a solution of hydrated protein and in the corresponding binary solution. Hence it is possible to calculate these activity coefficients from standard physicochemical data. It also enables us to understand in detail the equilibrium across a membrane separating a protein solution from a multi-component solution of low molecular weight substances.

With this thermodynamic understanding, it is possible to measure the water released when tobacco mosaic virus protein polymerizes, by using a sensitive spring-balance. Protein dissolved in a solution of water, glycerol and buffer is placed inside a cellophane sac suspended on a sensitive quartz helix. This sac is suspended in a solution containing water, glycerol and buffer. When the *pH* of the whole system, inside and outside the sac, is around 7.5, the protein is essentially unpolymerized; when the *pH* is reduced to 5.5 by replacing the external solution and allowing equilibrium to be established, the protein polymerizes. A reversible change in weight is detected by a small change in the extention of the quartz helix, previously calibrated. When polymerization is brought about in this manner, 0.029 g of water are released for each gram of protein. Polymerization can also be brought about by changing temperature at constant *pH*. A similar but somewhat more complex measurement with the spring-balance showed that 0.033 g of water were released per gram of protein upon polymerization brought about in this manner. The spring-balance experiments can also be used to determine the change in the amount of salt binding during the polymerization process. This was found to be very small or negligible.

Necessary for the interpretation of the spring-balance experiments, but also of interest of its own right, is the change in partial specific volume of protein during polymerization. This has been measured in a Linderström-Lang dilatometer. For a hypothetical polymerization experiment in which polymerization is brought about by changing *pH* from 7.6 to 5.5 without changing protein concentration, the increase in the partial specific volume of the protein can be calculated to be 0.00787 ml/g. Part of this increment must be attributed to hydrogen ion binding. This was estimated by making similar measurements on tobacco mosaic virus and was found to be 0.0013 ml/g. The balance of the partial specific volume of tobacco mosaic virus protein must be attributable to something other than hydrogen ion binding. It is interpreted as being the result of an increase in volume when water molecules in an ordered and somewhat compact state surrounding a protein molecule are released to become less compact free water. This increase in volume accompanying polymerization means that the application of hydrostatic pressure to a system in which polymerized and unpolymerized tobacco mosaic virus protein particles are in equilibrium would result in a shift of the equilibrium in the direction of depolymerization.

It seems reasonable to generalize and conclude that, since release of water molecules provides the entropy increase necessary for the spontaneous polymerization of tobacco mosaic virus protein, a similar release of water molecules provides the increase in entropy required for all, or most, entropy-driven biological and biochemical structure-forming processes, processes which resemble in many aspects of their behavior the polymerization of tobacco mosaic virus protein.

Chapter 3

Tobacco Mosaic Virus and Its Protein: Disc-Forming Polymerization

Introduction

STANLEY (1) purified tobacco mosaic virus (TMV) in paracrystalline form in 1935. The virus was soon thereafter shown to be a nucleoprotein containing 5% ribonucleic acid (RNA) (2,3,4), and to have a very high molecular weight (5,6). Double refraction of flow measurements first reported by TAKASAHI and RAWLINS (7) and refined by LAUFFER and STANLEY (8) led to the conclusion that the TMV particle is rod-shaped. From viscosity and sedimentation data LAUFFER (9) concluded in 1938 that the TMV particle is a long thin rod with a molecular weight of 42.5 million, thereby describing the general size and shape before the first electron micrographs were published. Subsequent studies made by many investigators, involving sedimentation, diffusion, viscosity, electron microscopy, x-ray diffraction and light-scattering confirmed and refined this conclusion. It is now generally agreed that the TMV particle is a rod, approximately 3,000 Å in length and 149 Å in "effective" diameter[1], with a molecular weight of 39 ± 1.2 million. From x-ray diffraction results (10,11) and electron microscopy (12, 13, 14) it has been established that the virus particle is a right-handed helix composed of more than two thousand protein subunits, intermeshed with a helically coiled strand of RNA with a molecular weight of about two million. The detailed structure for the virus particle has already been illustrated in Fig. 3 of Chap. 2. Further information is given in the review by LAUFFER and STEVENS (15).

SCHRAMM (16) found that when TMV is treated with dilute alkali, the protein component can be isolated in pure form. Because of this original method of preparation by treatment with alkali, the isolated protein is frequently

---

1 Because of pits and peaks in the surface of the TMV particle, illustrated in Fig. 3 of Chapter 2, the maximum diameter from peak to peak is about 190 Å.

called A-protein. An apparently identical protein can be isolated from virus degraded with 67% acetic acid (17). The protein subunit has the amino acid sequence shown in Fig. 1 (18), and a molecular weight of 17.53×10^3. The TMV model illustrated in Fig. 3 of Chap. 2, an interpretation of x-ray diffraction data, shows that the protein subunit in the virus must have one dimension equal to about 70 Å, and its maximum dimensions perpendicular to this must be about one third as great. A prolate ellipsoid of revolution with major semi axis of 35 Å and minor semi axis of 12.25 Å should come close to being a satisfactory model.

5 10 15 20
Acetyl-Ser-Tyr-Ser-Ile- Thr-Thr-Pro-Ser-GluN- Phe-Val-Phe-Leu-Ser-Ser-Ala-Try-Ala-Asp- Pro-Ile-Glu-Leu-Ile-

25 30 35 40
AspN-Leu-CysH-Thr-AspN- Ala-Leu-Gly-AspN-GluN- Phe-GluN-Thr-GluN-GluN- Ala-Arg-Thr-Val-Val-

45 50 55 60
GluN-Arg-GluN-Phe-Ser- GluN-Val-Try-Lys-Pro- Ser-Pro-GluN-Val-Thr- Val-Arg-Phe-Pro-Asp-

65 70 75 80
Ser-Asp-Phe-Lys-Val Tyr-Arg-Tyr-AspN-Ala- Val-Leu-Asp-Pro-Leu- Val-Thr-Ala-Leu-Leu-

85 90 95 100
Gly-Ala-Phe-Asp-Thr- Arg-AspN-Arg-Ile-Ile- Glu-Val-Glu-AspN-GluN- Ala-AspN-Pro-Thr-Thr-

105 110 115 120
Ala-Glu-Thr-Leu-Asp- Ala-Thr-Arg-Arg-Val- Asp-Asp-Ala-Thr-Val- Ala-Ile-Arg-Ser-Ala-

125 130 135 140
Ile-AspN-AspN-Leu-Ile- Val-Glu-Leu-Ile-Arg- Gly-Thr-Gly-Ser-Tyr- AspN-Arg-Ser-Ser-Phe-

145 150 155
Glu-Ser-Ser-Ser-Gly- Leu-Val-Try-Thr-Ser- Gly-Pro-Ala-Thr.

Fig. 1. Amino acid sequence of TMVP. From ANDERER *et al.* (18). Reprinted by permission of the copyright owner, Zeitschrift für Naturforschung

When tobacco mosaic virus protein (TMVP) is dissolved at a concentration of about 1 milligram per milliliter in approximately 0.1 ionic strength (μ) buffers at *pH* values between 6.5 and 8, it exists in a partially polymerized state. Osmotic pressure results of BANERJEE and LAUFFER (19) shown in Fig. 2, when extrapolated to zero concentration, yield a value of π/c corresponding to a number average molecular weight of approximately 50,000. This suggests that the predominant particle is a trimer of the chemical subunit with a molecular weight of 52.59×10^3. Since osmotic pressure is a measure of number average molecular weight, the simultaneous presence of both smaller and larger particles in equilibrium with the trimer is consistent with the observation.

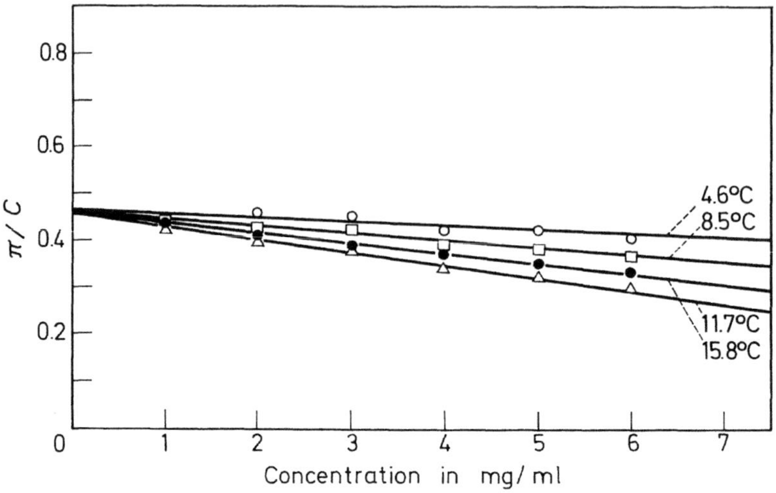

Fig. 2. Effect of temperature on osmotic pressure of TMVP in 0.066 μ phosphate buffer, *pH* 7.5, π is the osmotic pressure in centimeters of water, and *c* the protein concentration in milligrams per milliliter. Similar data were obtained with 0.1 μ, *pH*=6.5 phosphate buffer. From BANERJEE and LAUFFER (19). Reprinted by permission of the copyright owner, The American Chemical Society

CASPAR (20) proposed on theoretical grounds that a cyclical trimer represents a stable intermediate in the polymerization process. In addition to that presented above and presented in some detail by LAUFFER and STEVENS (15), supporting evidence for its existence comes from the fact that the sedimentation coefficient of TMVP in solutions comparable to the ones under discussion has approximately the right value for a cyclical trimer. In the cyclical trimer, it is imagined that the three chemical subunits are arranged with their long dimensions approximately parallel and are stacked together such that each touches the other two laterally near the equator. In cross-section, the centers of the three monomers form the apeces of an equilateral triangle.

Dissociation of Trimer

In 1959 ANSEVIN and LAUFFER (21) showed that, at temperatures between 0° and $6^{o}C$, TMVP with a sedimentation coefficient of around 4 Svedbergs dissociates into material with a sedimentation coefficient of 1.9 Svedbergs when the concentration is reduced from several mg/ml to 0.1 mg/ml. The data from their experiment are shown in Fig. 3. Hydrodynamic calculations show that 4 Svedbergs is approximately the right sedimentation coefficient for a cyclical trimer, and 1.9 Svedbergs is approximately correct for the monomeric protein subunit.

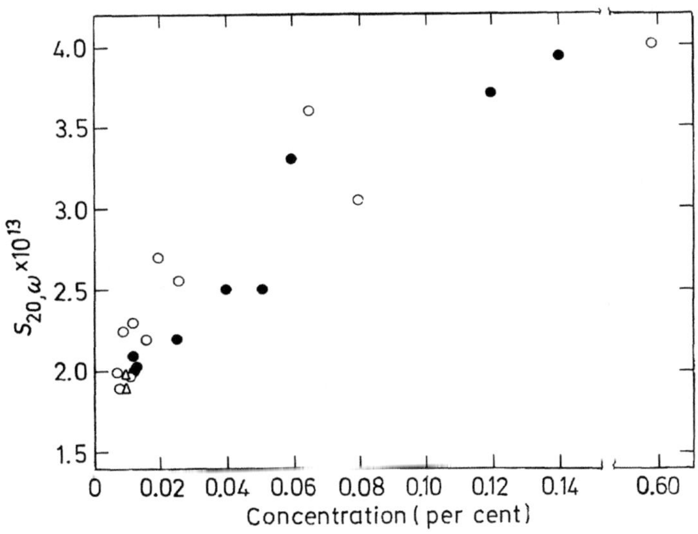

Fig. 3. Sedimentation coefficients of TMVP at various concentrations. ○, 0.1 μ borate buffer, *pH*=9.0; Δ, 0.05 μ phosphate buffer, *pH*=7.0; ●, 0.10 μ phosphate buffer, *pH*=6.5. From ANSEVIN and LAUFFER (21). Reprinted by permission of the copyright owner, Macmillan (Journals) Ltd.

It can be observed in Fig. 3 that at a concentration of 0.5 mg/ml, the sedimentation coefficient is about half way between its highest and its lowest values. It is reasonable to suppose that this represents an unresolved mixture in which half of the protein is sedimenting with a coefficient of 4 Svedbergs, and the other half with a coefficient of 1.9 Svedbergs. It is possible to interpret this analysis in terms of the equilibrium, $3S \rightleftarrows S_3$, where S means the chemical subunit and S_3 means the cyclical trimer. The mass-action equation for this reaction is

$$K = [S_3]/[S]^3 = (f)/3m^2(1 - f)^3 \qquad (1)$$

In this equation, the brackets signify molar concentration, m means the molarity of total protein calculated as chemical subunit and f is the weight fraction polymerized. If f is 0.5 when m is 0.5/17,500 or 0.285×10^{-4}, K has a value of 0.16×10^{10}. With the aid of Eq. (2) of Chap. 1, one can readily calculate that the standard free energy change for converting 3 moles of monomer into trimer is slightly less than -12,000 cal. Similar calculations with about the same result were reported by CASPAR (20). WESTOVER and STEVENS, as reported by LAUFFER and STEVENS (15), carried out a more detailed analysis by means of sedimentation equilibrium. For solutions in 0.1 M KCl and 0.001 M phosphate buffer at pH=7.1, their results are consistent with the view that the main components are monomer in equilibrium with trimer. The standard free energy for the formation of 1 mole of trimer from 3 moles of chemical subunit was found to be -12,500 cal. DURHAM and KLUG (22) estimated from sedimentation equilibrium studies a value for the standard free energy change of -15,00 cal. per mole of trimer formed from three monomers. However, they defined the standard state as unit mole fraction. When their result is recalculation to a standard state of unit molarity[2], the value for the standard free energy

---

[2] For a bimolecular reaction,

$$K_m = \frac{m_2}{(m_1)^2} = \frac{55.6\ X_2}{(55.6\ X_1)^2} = \frac{1}{55.6} K_x, \text{ and}$$

$$- RT \ln K_m = \Delta F_m^o = 3.39\ RT + \Delta F_x{}^o.$$

Similarly, for a trimolecular reaction,

$$\Delta F_m^o = 6.78\ RT + \Delta F_x.$$

X and m mean mole fraction and molarity, respectively, and the subscripts x and m denote standard states of unit mole fraction and unit molarity, respectively.

change becomes -11,300 cal/mole of trimer, in reasonable agreement with the previously mentioned figures.

The protein subunit with molcular weight 17,500 is also obtained in 67% acetic acid (19) (23) and in alkaline solutions around *pH* 13, (24,25). When the protein from cucumber virus 4, a virus serologically and morphologically similar to TMV, is treated with succinic anhydride, the protein dissociates into particles with a molecular weight close to 17,500 (26). TMVP is less completely dissociated by reaction with succinic anhydride.

General Preview of the Polymerization of TMV Protein

The polymerization of TMVP, assumed to exist essentially as a trimer with molecular weight of approximately 52,500 when dissolved in 0.1 μ phosphate buffer at *pH*=6.5, has been studied extensively in the author's laboratory over a period of more than a decade. In 1967 it was pointed out that the sometimes conflicting results of these studies could be understood in terms of the assumption that there are two radically different processes involved in TMVP polymerization (27). One of these takes place at temperatures between 4 and about 12°C when protein at a concentration of 1 milligram per milliliter is dissolved in 0.1 μ phosphate buffer at *pH*=6.5. The other takes place at temperatures above 12 or 14°C when the protein is dissolved in the same medium. The first process was designated "low-temperature" polymerization and the second process, "high-temperature" polymerization. The names "low-temperature" polymerization and "high-temperature" polymerization are of historical significance only. Actually, the process which takes place at "low-temperature" at *pH*=6.5 takes place at both low and high temperatures (near room temperature) at *pH*=7.5, and the process described as high-temperature polymerization takes place at temperatures well under 12°C when the *pH* is 6.0 or lower. In view of evidence accumulated since 1967, it is preferable to describe these two processes as disc-forming and rod-forming polymerization, respectively.

To orient the reader for the balance of this chapter and for Chaps. 4 and 5, a preview of our current understanding of the various kinds of polymerization exhibited by TMVP is here presented. The mechanism of polymerization of the proteins from TMV and several of its strains can be summarized by the scheme shown below. The symbol, A, means relatively unpolymerized protein such as that found at $4^{o}C$ in a 1 mg/ml solution in 0.1 μ phosphate buffer at *pH*=6.5 with a number average molecular weight of approximately 50,000; the question of whether A means homogeneous cyclical trimer or an equilibrium mixture of larger and smaller particles is left open. The symbol, R, means the short rod or double spiral as described in Chap. 4, D means double discs such as those described in this Chapter. P means rod-like polymers in which the protein subunits are arranged helically as they are in the virus itself, and S means stacked double discs. Drawings of the various polymers are shown in Fig. 4. H^{+} means the hydrogen ion. The symbols, *n*, *m*, *x*, *y* and *z*, are merely the numbers required to make the equations balance. Solid arrows represent experi-

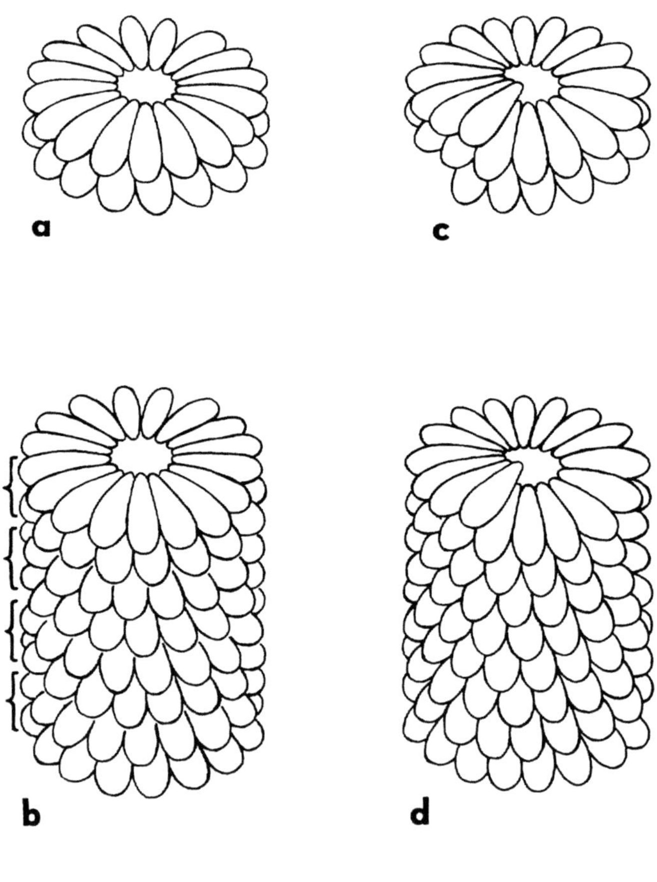

Fig. 4 a-d. Models of four aggregates of tobacco mosaic virus protein. (a) The double disc; (b) The stacked double disc (brackets identify individual double discs); (c) The hypothetical double spiral; (d) The helical rod.

mentally established reactions, and dotted arrows correspond to possible reactions.

$$nA + xH^+ \overset{I}{\rightleftharpoons} R \qquad ; \qquad mR + yH^+ \overset{II}{\rightleftharpoons} P$$

$$R \overset{IV}{\rightleftharpoons} (x-z)H^+ + D \qquad ; \qquad P \overset{VI}{\rightleftharpoons} (mx+y)H^+ + S$$

$$nA + zH^+ \overset{III}{\rightleftharpoons} D \qquad ; \qquad mD \overset{V}{\rightleftharpoons} S$$

The findings leading to this view are discussed in Chaps. 3, 4 and 5. The remainder of this Chapter will be confined to a discussion of disc-forming polymerization, Reaction III and stacking of double discs, Reaction V.

Disc-Forming Polymerization of TMV Protein

The negative slopes of the π/c vs c plots found in Fig. 2 show that as concentration increase, the fraction of larger polymers in the equilibrium mixture increases. The fact that the negative slope is greater the higher the temperature indicates that the extent of polymerization increases with increasing temperature, thus identifying this polymerization as an endothermic process. The data shown in Fig. 2 were obtained for 0.066 μ phosphate buffer at pH=7.5. Similar results were obtained at temperatures between about 4.0 and 12.8°C for the protein dissolved in 0.1 μ phosphate buffer at pH=6.5 (19, 28). LAUFFER and co-workers (27) reported that electron micrographs obtained after spraying solutions of TMVP at concentrations of 6.5 and 9 mg/ml in 0.1 ionic strength phosphate-ammonium acetate buffer at pH=7.5 onto electron microscope grids at 20°C showed predominately disc-like structures. These are illustrated in Fig. 5. They were first described in detail by MARKHAM and co-workers (29), who obtained evidence by means of electron microscopy that TMVP polymerizes under some circumstances in the form of a double disc with a hole in the center, each member of the pair of discs being made up of 16 protein monomers. Previous evidence for the existence of such discs was available as early as 1955 from X-ray diffraction studies by FRANKLIN and COMMONER (30). More recently, FINCH and co-workers (31) obtained evidence from x-ray diffraction and electron microscopy of similar double discs, but with 17 protein monomers

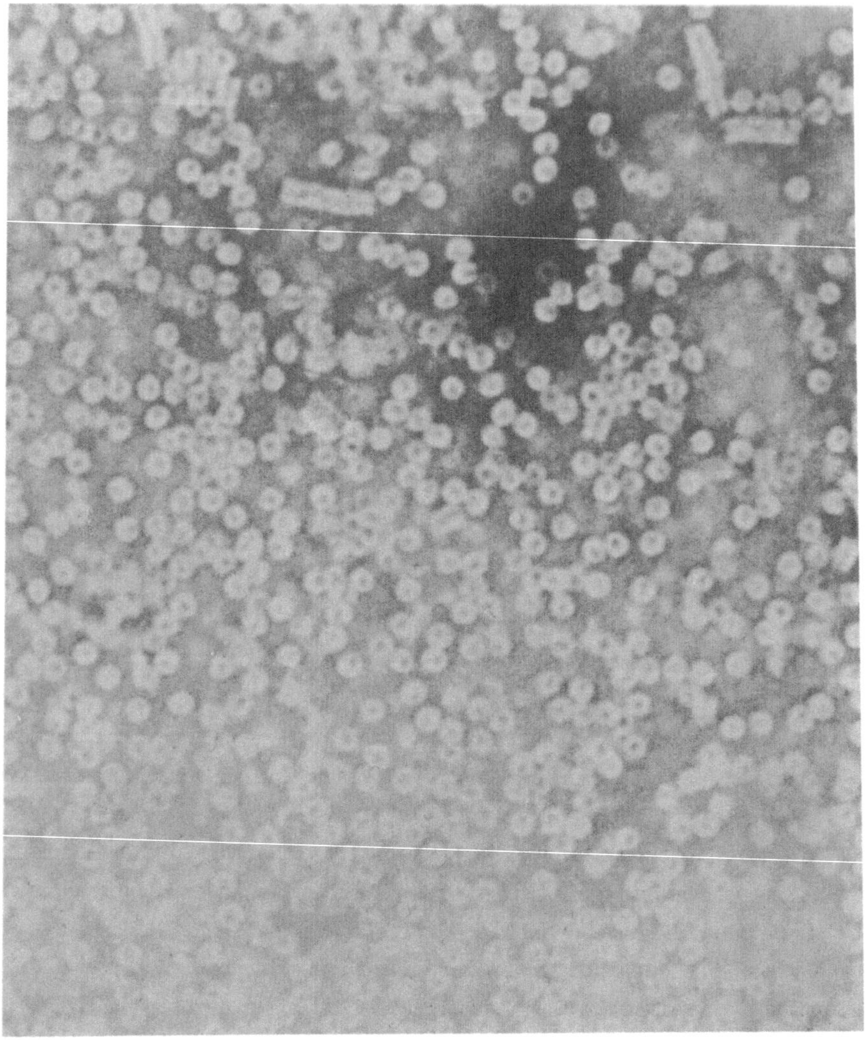

Fig. 5. TMVP at 6.5 mg/ml in 0.1 μ phosphate-ammonium acetate buffer, *pH*=7.5, 20°C. Magnification: about X 165,000. From LAUFFER and STEVENS (15). Reprinted by permission of the copyright owner, Academic Press, Inc.

in each ring. It is probable that double discs exist in solution as well as in the dry state necessary for electron microscopy and X-ray diffraction experiments. Components with sedimentation coefficients in the neighborhood of 20 Svedbergs have been observed on numerous occasions. This is approximately the right sedimentation coefficient for a double disc composed of 34 monomeric units (20). It is evident that this polymerization leads to the formation of double discs. This is a basically different process from the one already discussed in a preliminary way in Chapter 2 which leads ultimately to the formation of rod-like helices, a process which can be identified with what was called "high-temperature" polymerization, and will called rod-forming polymerization.

Linear Condensation Polymerization Mathematics

The early stage of polymerization of TMVP trimers can be interpreted in terms of the mathematics of Flory's theory of linear condensation polymerization (32), a theory based on the assumption that the chemical reactivity of a combining site on a unit in a chain is completely independent of the length of the chain of which it is the end. Imagine one milliliter of a solution containing N particles of all degrees of polymerization, N_x particles composed of x units each in a linear array (curved in this instance), and N_0 total units, including free units and those combined in linear polymer particles. Assume that, on one end of each uncombined unit and on one end of each polymer, a particular kind of combining site is free. On all other units, that is on all units within a polymer chain except the terminal one, this site will be combined. If p is the probability that such a site is combined, $(1-p)$ will be the probability that it is free. Therefore, in the solution the number of combined sites will be N_0p, and the number of free sites will be $N_0(1-p)$. Since each particle, whether an uncombined unit or a polymer particle, will have one free site, $N_0(1-p)$ will also be the number of particles, N. The probability of finding a polymer composed of x units, N_x/N, is the product of the probability of finding in sequence $(x-1)$ combined sites, p^{x-1}, and the probability of finding one uncombined site, $(1-p)$. Thus, one can write

$$N_x/N = (1-p)\ p^{(x-1)} \tag{2}$$

Since N is equal to $N_0(1-p)$,

$$N_x/N_o = (1-p)^2\ p^{(x-1)} \tag{3}$$

For a hypothetical biomolecular equilibrium reaction, $P_1 + P_1 = P_2$, one can write from the mass-action law;

$$[P_2]/[P_1] = K[P_1] \tag{4}$$

In this equation, K is the equilibrium constant, and the brackets indicate equilibrium concentrations. If the combining sites on P_1 and P_2 have the same reactivity, then exactly the same equilibrium constant should apply to the reaction $P_2 + P_1 = P_3$. Thus,

$$[P_3]/[P_2] = K[P_1] \tag{5}$$

When the corresponding terms of Eqs. (4) and (5) are multiplied,

$$[P_3]/[P_1] = (K[P_1])^2 = (K[P_1])^{3-1} \tag{6}$$

This argument can be extended to give the general equation,

$$[P_x]/[P_1] = (K[P_1])^{(x-1)} = N_x/N_1 \tag{7}$$

From Eq. (3) it follows that $N_1/N_o = (1-p)^2$ and therefore N_x/N_1, which is the same as $[P_x]/[P_1]$, is $p^{(x-1)}$. Therefore, p equals $K[P_1]$ and, since $[P_1]/[P_o]$ $= N_1/N_o = (1-p)^2$,

$$p = K\,[P]_o(1-p)^2 \quad \text{or} \quad K = p/[P]_o(1-p)^2 \tag{8}$$

It is important to observe that the K of equation 8 is the simple second order equilibrium constant for the reaction $P_1 + P_1 = P_2$ and all similar reactions involving P_1 or polymers of P_1.

Osmotic Pressure Studies

In the limiting case, osmotic pressure depends on the number of particles per unit volume. Therefore, for a solution which is ideal except for the tendency of P to polymerize, it follows by dividing Eq. (3) by Eq. (2) that

$$N/N_o = \pi/\pi_o = (1-p). \tag{9}$$

In this equation, π is the osmotic pressure of a solution containing polymer chains of various lengths in equilibrium, and π_0 is the osmotic pressure the same solution would have if all of the material were unpolymerized. When Eq. (9) is solved for p and is substituted into Eq. (8), Eq. (10) is obtained.

$$\pi = \frac{\pi_0^2}{K[P]_0} \frac{1}{\pi} - \frac{\pi_0}{K[P]_0} \qquad (10)$$

Since $\pi_0 = RT\,[P]_0 = 1{,}000\ RT\ c/M_0$, this equation can be transformed into

$$\pi = \frac{1000(RT)^2}{M_0 K} \frac{c}{\pi} - \frac{RT}{K} \qquad (11)$$

In Eq. (11), R is the gas constant, T the absolute temperature, c the total protein concentration in grams per milliliter, and M_0 the molecular weight of the individual repeat unit in the polymer chains. This equation states that if the early stage of polymerization of TMVP adhears to Flory's formulation, one should get a straight line if π is plotted against c/π, and then from the intercept and the slope of that line one can evaluate both K and M_0.

This is admittedly an unusual way to plot osmotic pressure data. However, it must be remembered that it is intended only for systems which polymerize in accordance with the assumptions made by Flory, and which are dilute enough so that they can be considered to be ideal except for the tendency to polymerize. When data obtained by BANERJEE and LAUFFER (19) for polymerization of TMVP in 0.1 μ phosphate buffer at pH=6.5 are plotted in this manner, they do fit straight lines, as shown in Fig. 6. The slopes and the intercepts of the lines depend upon temperature. Table 1 shows the values of M_0 and K evaluated with the aid of equation (11) for these data at three different temperatures. It is evident that M_0 is close to 52,500, the molecular weight of the cyclical trimer of TMVP. This is perhaps the best evidence we have that the cyclical trimer is indeed the polymerizing unit in early state polymerization.

PAGLINI and LAUFFER (28) extended studies of this sort to include investigation of TMVP polymerization in 0.1 μ barbital buffers at pH=7.5, and measurements in D_2O. The studies of PAGLINI and LAUFFER involved a much wider range of protein concentrations than those of BANERJEE and LAUFFER (19). It was found that data obtained in the phosphate and barbital buffers at $4^{o}C$ fit this same kind of plot up to protein concentrations of about 20 mg/ml, but

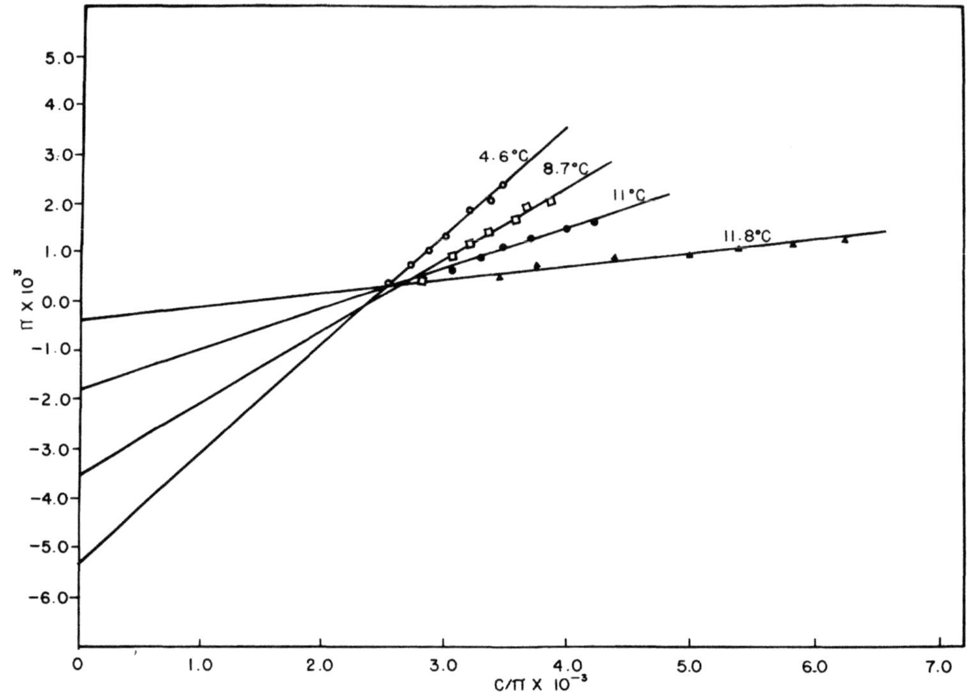

Fig. 6. Plot of π *vs.* c/π for TMVP in 0.1 μ phosphate buffer, *pH*=6.5, as a function of temperature; π is osmotic pressure in centimeters of water, c is protein concentration in milligrams per milliliter. Data of BANERJEE and LAUFFER (19). From LAUFFER and STEVENS (15). Reprinted by permission of the copyright owner, Academic Press, Inc.

that at higher protein concentrations significant deviations were observed. PAGLINI and LAUFFER presented a theory for correcting the data obtained at the higher protein concentration - a theory based on hypothetical contributions to non-ideality arising from electrical charge (Donnan effect), and the excluded volume effect. With such corrections, data obtained at concentrations up to about 30 mg/ml fit the theory. Based on calculations made on the assumption that M_0 is 52,500 using Eq. (11) properly modified, PAGLINI and LAUFFER obtained the values of K shown in Table 2. The values of $\overline{M}_n$ in Table 2 were obtained by extrapolating data in the manner shown in Fig. 2.

It is evident that the osmotic pressure studies discussed above can be interpreted successfully in terms of Flory's theory of condensation polymerization.

Table 1. Variation of equilibrium constant with temperature for disc-forming polymerization of TMVP

T(^{o}C)	K	M_o	$-\Delta F^o$ (kcal/mole)
4.6	4.21 x 10^3	54,000	4.6
8.7	6.99 x 10^3	53,000	4.9
11.0	12.93 x 10^3	50,500	5.3

Both equilibrium constants, K, and number average molecular weights, M_n, were evaluated by fitting the data to Eq. (11). From BANERJEE and LAUFFER (19). Reprinted by permission of the copyright owner, The American Chemical Society.

DURHAM and KLUG (22) arrived at a similar conclusion from analysis of sedimentation equilibrium experiments. They state "the polymerization process thus has a close affinity with condensation polymerization..." As was first pointed out by PAGLINI and LAUFFER (28), it is quite surprising that a polymerization process which leads to the formation of discs or closed rings should proceed, even in its early stages, according to the pattern of linear condensation polymerization. The reason is that the standard free energy for the formation of a union, related to the equilibrium constant as shown in Eq. (2) of Chap.1, should be the same for each union. But when the final unit which closes a ring is added, two unions are formed. If the standard free energy decrease for adding this final unit is twice that for adding other units, because twice as many unions are formed, then the equilibrium should be shifted sharply in the direction of discs, with the resultant depletion of intermediate aggregates. PAGLINI and LAUFFER indicated that this distortion should be observable when π/π_o has values no lower than 0.7. This distortion should lead to values of π smaller than those predicted by simple condensation polymerization mathematics and therefore to significant deviations from Eq. (11). Since such deviations do not occur, one is forced to conclude that the ring closing step does not have twice the standard free energy decrease of the simple linear addition steps. A simple way to explain this is to assume that there is strain associated with the final step. This idea has been developed quantitatively by the author (33), who showed that a value of 3.5 kcal per mole for the free energy of strain would suffice.

Table 2. Variation of equilibrium constant with temperature for disc-forming polymerization of TMVP[a]

Buffer In H_2O	*pH*	Temp(^oC)	Ionic Strength	$\overline{M}_n$	*K*
PO_4	6.5	4	0.1	50,000	5.56×10^3
PO_4	6.5	11.7	0.1	64,155	29.2×10^3
PO_4	7.5	4	0.1	53,410	4.14×10^3
PO_4	7.5	20	0.1	100,200	27.2×10^3
PO_4	7.5	25	0.1	101,104	33.3×10^3
Barbital	6.5	4	0.1	50,865	3.26×10^3
Barbital	6.5	4	0.055	48,755	2.06×10^3
Barbital	7.5	4	0.1	50,213	3.04×10^3
In D_2O	*pD*				
PO_4	7.0	4	0.1	141,720	75.7×10^3
PO_4	8.0	4	0.1	75,920	11.2×10^3

[a] In this table, in contrast with the procedure used for the data in Table 1, the values of $\overline{M}_n$ were obtained by extrapolating the experimental data to zero concentration in the manner illustrated in Fig. 2. The values for the equilibrium constants, *K*, however, were obtained from the best possible fit of the experimental data to Eq. (11), on the assumption that M_o is 52,500. Condensed from PAGLINI and LAUFFER (28). Reprinted by permission of the copyright owner, The American Chemical Society.

As was explained in Chap. 1, the van't Hoff enthalpy and the standard entropy change can be evaluated by plotting ln *K* against $1/T$ in accordance with Eq. (3) of Chap. 1. A straight line should be obtained with a slope of $-\Delta H^o/R$ and an intercept of $\Delta S^o/R$. Such a plot of data of BANERJEE and LAUFFER similar to those shown in Table 1 is presented in Fig. 7. From the slope and the intercept one can calculate that ΔH^o is +30,000 cal/mol, and ΔS^o is +124 entropy units for disc-forming polymerization of TMV protein in 0.1 μ phosphate buffer at *pH*=6.5. PAGLINI and LAUFFER (28) estimated values of +34,00 cal/mol, +139 entropy units for polymerization 0.1 μ phosphate buffer at *pH*=6.5, and of +19,000 cal/mol, +86 entropy units for 0.1 μ phosphate buffer at *pH*=7.5.

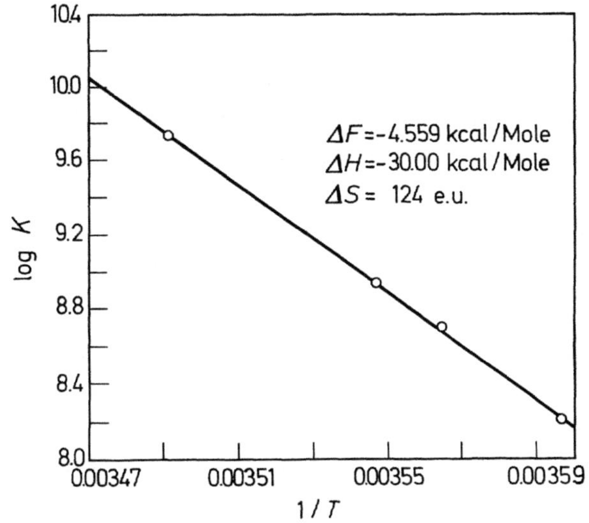

Fig. 7. Plot of ln *K* vs. $1/T$, where *K* is the equilibrium constant and *T* is the absolute temperature. From BANERJEE and LAUFFER (19). Reprinted by permission of the copyright owner, The American Chemical Society

It is quite evident that the theory that the formation of double discs proceeds by a condensation polymerization mechanism with a trimer of molecular weight 52,500 as the basic repeat unit fits osmotic pressure data well at *pH* values of 6.5 and above and at temperatures below about 10°C. This is, however, true only at protein concentrations above 1 or 2 mg/ml. It is well known that trimers dissociate into monomers at concentrations below approximately 1 mg/ml. DURHAM and KLUG (34) report that, when allowance is made for dissociation of trimers into monomers at very low concentrations, this theory also fits sedimentation equilibrium data. However, they reject this theory "because there is no structural reason for trimers to retain their identity inside a larger polymer and for a hexamer, for example, to be allowed but not a pentamer." This reason seems to represent a misunderstanding of the theory. In the actual situation, every conceivable equilibrium reaction can occur. It is taken for granted that the state of the system is dynamic; otherwise it would be folly to apply equations based on equilibrium thermodynamics. There is no reason inherent in this theory why a hexamer cannot dissociate a monomer to become a pentamer for a fleeting moment or combine with a stray monomer to form a heptamer. The theory predicts, however, that a sufficiently accurate description of the number average molecular weight of the system can be obtained by focusing attention on the equilibrium between trimers and integral multiples of trimers. It is obvious that one could never get a double disc solely by the polymerization of trimers or integral multiples of trimers; the double disc contains a number of monomers which is not a multiple of three.

DURHAM and KLUG (34) proposed a modified version of the condensation polymerization theory. In this version, monomers are in equilibrium with trimers, which form nuclei essential for further polymerization. This further polymerization takes place by adding monomers. When proper allowance is made for strain, this theory also fits their sedimentation equilibrium data. There is little doubt that this theory would also fit osmotic pressure data because of its extra equilibrium constant, which can be assigned a value to make it fit. It is unlikely that curve fitting exercises can differentiate between the two theories. For the present, at least, the decision between them must be largely aesthetic. To those for whom the principle of parsimony has strong appeal, Lauffer's theory will seem preferable because it involves only one arbitrarily assigned equilibrium constant, in contrast with two for the other. One important fact must be remembered: whichever equilibrium process is postulated, the thermodynamic parameters evaluated by data fitting must correspond to the type of equilibrium assumed. In Lauffer's theory the values of ΔH° and ΔS° correspond to the addition of a trimer or larger polymer to a trimer or larger polymer. In the theory of DURHAM and KLUG, the thermodynamic parameters refer to the addition of a monomer to a trimer or to a larger polymer.

In the theory of DURHAM and KLUG, it is difficult to see why the free energy of adding a monomer to a dimer should differ significantly from the free energy of adding a monomer to a trimer or higher polymer, because in both cases two contacts are made between the monomer and the larger particle. Thus, there is no apparent reason why a dimer should not be the seed particle rather than a trimer. The two theories have one important feature in common. They are both condensation polymerization theories, theories which involve a sequence of larger and larger particles, all simultaneously in equilibrium. Thus they are distinctly different from theories which postulate the coexistence exclusively of end products (double discs in this case) and initial products (monomers and/or trimers). It is generally agreed that such theories can be eliminated.

Calorimetric Studies

STAUFFER, *et al.* (35) measured calorimetrically the endothermic enthalpy of disc-forming polymerization of TMVP by diluting TMVP at a concentration of 50 mg/ml in 0.1 ionic strength phosphate buffer at pH=7.5 to various concentrations of protein in (a) 0.1 μ phosphate buffer at pH=7.5, (b) in 0.1 μ phosphate buffer at pH=8, and (c) in 0.01 μ phosphate buffer at pH=7.5. Since this is a process which involves depolymerization, the heats of polymerization are opposite in sign to the observed heats. In all cases, heats associated with changing media were subtracted out in parallel controlled experiments. Some of the experiments involved several steps, and in these cases the overall heats were calculated using Hess's Law.

Calorimetrically measured heat divided by the mass of protein gives the heat released or absorbed per gram of protein. In order to compare these results with van't Hoff enthalpies, they must be converted to molar quantities. If M_1 and M_2 represent the number average molecular weight in the partially polymerized and in the highly polymerized states, respectively, then the number of molecules per gram of protein is A/M_1 for the partially polymerized and A/M_2 for the highly polymerized state, respectively. A is Avagadro's number. When the molecules in both the partially unpolymerized and the highly polymerized state are linear, then the number of free ends per gram is the same as the number of molecules per gram. The enthalpy per mole of "bond" formed is then equal to Avagadro's number times the heat per gram, q, divided by the change in the number of free ends per gram. This is given by Eq. (12).

$$\Delta H = \frac{Aq}{(A/M_1)} - \frac{Aq}{(A/M_2)} = \frac{q}{(1/M_1 - 1/M_2)} \qquad (12)$$

For the present case, however, the best evidence available indicates that the highly polymerized particles found at a concentration of 50 mg/ml in 0.1 μ phosphate buffer at pH=7.5 are in the form of double-discs. In such a case, the number of free ends in state 2, the highly polymerized state, is zero and the number of free ends in state 1 is $A(1 - r)/M_1$, where r is the ratio of the number of discs to the total number of molecules. In this case,

$$\Delta H = \frac{Aq}{A(1 - r)/M_1} - 0 = q\, M_1/(1 - r) \qquad (13)$$

Ordinarily, it is difficult to determine the value of r. However, it is a parameter which should have a value of one at very high protein concentration and of zero as the protein concentration approaches zero and therefore exists in the minimal state of polymerization. Thus, one should be able to evaluate ΔH by measuring both q and M_1 for a variety of protein concentrations in the dilute state, followed by extrapolation of $q\, M_1$ to zero protein concentration. The values of M_1 for the relatively unpolymerized material were obtained by osmometry. The results of such experiments are shown in Fig. 8. The extrapolated values fall within the range of 25,000 to 30,000 cal/mole of "bond". These are in reasonably good agreement with the van't Hoff enthalpies obtained by BANERJEE and LAUFFER (19) and by PAGLINI and LAUFFER (28). However, as was pointed out by STAUFFER, *et al.*, the equilibrium measurements refer to ΔH values for the initial stage of the process of polymerization in which a negligible fraction of the material is in the double-disc state. The values obtained in the above manner from calorimetry, even when extrapolated to zero concentration, are average enthalpies per mole of "bond" for all "bonds", including the final one when discs are closed. It is reasonable to assume that the thermodynamic parameters are the same for all successive stages in the polymerization process short of the final ring closing step, but it is probable that the enthalpy of this step is higher, as a result of the strain discussed previously. Therefore, the values obtained by calorimetry, even when treated as above, should never be precisely the same as van't Hoff enthalpies. There is the further problem, summarized by Eq. (5) of Chap. 1, that van't Hoff enthalpies can equal calorimetric enthalpies only when ΔC_p, the difference between the heat capacity of products and reactants, is equal to zero. We have no information about ΔC_p.

While the extrapolated values of enthalpy per mole obtained by calorimetry, shown in Fig. 8, are in reasonable agreement with the van't Hoff enthalpies obtained by BANERJEE and LAUFFER (19) and PAGLINI and LAUFFER (28) calculated on the assumption that the polymerizing unit is the cyclical trimer, they are much higher than the van't Hoff enthalpy reported by DURHAM and KLUG (34), 14.6 kcal/mole, calculated on the assumption that the polymerization involves the addition of monomers. While this result seems to weigh somewhat in favor of the former theory over the latter, a note of caution must be introduced. The extrapolated values shown in Fig. 8 are probably reliable for the two upper curves, because they do not differ greatly from the experimental values at 2 and 3 mg/ml. However, in the case of the lower curve, the extrapolation is less reliable because of the large increment beyond the experimental values. The possibility, therefore, remains that a "correctly" extrapolated value of the lower curve would come closer to the van't Hoff enthalpy calculated on the basis of the theory of DURHAM and KLUG. All in all, it is not possible with present knowledge to decide unambiguously between the two theories.

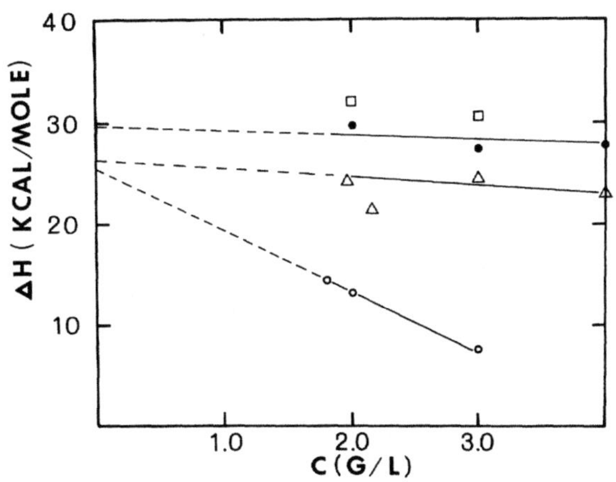

Fig. 8. Graph of ΔH as a function of final TMVP concentration. Initial state: 50 mg of protein per ml in 0.1 μ phosphate buffer, *pH*=7.5. ○, final state, 0.1 μ, *pH*=7.5; Δ, final state, 0.1 μ, *pH*=8.0; ●, final state, 0.01 μ *pH*=7.5. From STAUFFER *et al.* (35). Reprinted by permission of the copyright owner, The American Chemical Society

It is necessary to be quite clear about what one means by the word "bond". If the starting material is a double disc and one break is made in it, there will be no change in the molecular weight of the particle. However, in accordance with the theory of LAUFFER, when a partially completed double disc consisting of two layers of protein subunits forming a portion, perhaps resembling the letter c, of a ring is broken, two polymer particles will be formed from the original. The word "bond" is here used in such a sense that the process just described would involve breaking one "bond". It is clear that this definition fits the operations involved in the calorimetric studies, because the heats evolved were interpreted in terms of the average molecular weights of the lower molecular weight end-products, and, therefore, in terms of the numbers of particles per unit volume. In reality, when one breaks a segment of a double disc, one breaks three contacts between protein subunits, one lateral contact between two subunits in the top layer, one lateral contact between two subunits in the bottom layer and one diagonal contact between one subunit in the bottom layer and one subunit in the top layer. Thus, a "bond" involves three contacts between protein subunits. If the depolymerization of a partially formed disc is thought of as involving initially the removal of a single protein subunit from the end, as proposed by DURHAM and KLUG (34), the breaking of such a "bond" would involve the disruption of two subunit-subunit contacts.

From the equilibrium constants at three different temperatures for the polymerization of TMVP in 0.1 μ phosphate buffer at *pH*=6.5, listed in Table 1, it is easy to calculate the standard free energies for "bond" formation. This is done by using Eq. (2) of Chap. 1. The resultant values range from -4.6 to -5.3 kcal/mole, depending upon the temperature. These are presumed to involve three protein subunit-protein subunit contacts. The formation of a cyclical trimer from protein subunits also involves the establishment of three protein subunit-protein subunit contacts, but the free energy, as discussed previously, is about -12 kcal/mole. The bonding in trimer formation is vastly different from that in further polymerization.

Effect of pH and Ionic Strength

Disc-forming polymerization differs from rod-forming polymerization in a number of fundamental ways. Unlike rod-forming polymerization, the disc-forming process is relatively insensitive to changes in *pH*. The effect of increasing concentration on polymerization in 0.1 μ phosphate buffer at 4°C is only slightly greater at *pH*=6.5 than at *pH*=7.5 or 8.0, (19). SCHEELE and LAUFFER (36) showed that at 4°C TMVP binds little or no hydrogen ion when titrated from approximately *pH*=9 to approximately *pH*=6.5. The same result was obtained whether the protein concentration was 7 mg/ml or 3 mg/ml. Thus, it can be stated that essentially no hydrogen ions are bound or released when the concentration of protein is changed under these conditions. Since the results of Fig. 2 show that the state of polymerization changes when concentration changes, it follows that there is little or no change in hydrogen ion binding during disc-forming polymerization at *pH* values between 6.5 and 8.0. As will be shown in the next chapter, this is not true of rod-forming polymerization. The results just discussed imply that z in reaction III of the scheme presented earlier in this Chapter is a very small number. As will be shown in the next Chapter, the hydrogen ion titration curve of TMV protein in the virus and in the helical rod (P) polymerized form exhibits two hydrogen ion binding centers per protein monomers with pK values in the neighborhood of *pH* 6-7. These are assumed to be abnormal carboxyl groups. Hydrogen ion titration curves published by DURHAM and KLUG (22) on double discs show no such abnormally titrating groups. As will be shown later, neither does unpolymerized virus protein. The inference from this result is that z is actually zero. However, previously published results from our laboratory (35) had shown that at a concentration of about 50 mg/ml in 0.1 ionic strength phosphate buffer at *pH* 7.5 and at room temperature, double discs are the predominate form. At *pH* 8.0, smaller aggregates are found. This implies that z is greater than zero. Furthermore, VOGEL and JAENICKE (37) have demonstrated a good correlation between hydrogen ion binding at *pH* 7 and the shift in mean residue ellipticity at 280 mμ associated with the formation of double discs, which was mentioned in Chapter 1. Their results indicate that z has a value of one hydrogen ion per protein subunit. Furthermore, they postulate that x, the value of hydrogen ion uptake in reaction I, has a value of two hydrogen ions per subunit.

Calorimentric data of STAUFFER, *et al.* (35), show that when one gram of TMVP is transferred at room temperature from 0.01 μ *pH*=7.5 phosphate buffer to 0.1 μ *pH*=7.5 phosphate buffer, 0.065 cal are absorbed. Osmotic pressure data reported by the same workers show that at room temperature in phosphate buffers at *pH*=7.5, the number average molecular weight is almost 40% greater when the ionic strength is 0.1 than when it is 0.01, protein concentrations being the same. From Table 2, the data of PAGLINI and LAUFFER (28), it can be seen that the equilibrium constant for polymerization at $4^{\circ}C$ in *pH*=6.5 barbital buffer is greater at 0.1 μ than at 0.055 μ. All of the above results consistently show that the degree of polymerization increases when ionic strength is increased. DURHAM and co-workers arrived at the same conclusion (38). The data of Table 2 also show that substituting D_2O for H_2O strongly promotes disc-forming polymerization.

Volume Changes

Another respect in which disc-forming polymerization differs from rod-forming polymerization is in the volume changes involved. The data in Table 1 of Chap. 2 show that when 0.13 g of protein dissolved in 7 ml of 0.10 molar sodium cloride solution are diluted to a volume of 32 ml, the partial specific volume, V_2, decreases only 0.00046 ml/g. This involves a change in protein concentration from about 18.6 mg/ml to about 4.1 mg/ml, a change which should be accompanied by a substantial degree of depolymerization, as indicated by the negative slope of the π/c *vs* c data shown in Fig. 2.

By using the value of K shown in Table 2 for polymerization at $4^{\circ}C$ in 0.1 μ phosphate buffer at pH=7.5, 4.14×10^3, it is possible to estimate the degree of polymerization, p, at the higher and at the lower concentrations. When one takes the value 52,500 for the molecular weight of the polymerizing unit, concentrations of 4.1 and 18.6 mg/ml correspond to 0.78×10^{-4} and 3.5×10^{-4} mole/l, respectively. The corresponding values of p calculated by using Eq. (8) are 0.20 and 0.45, respectively. Thus, the change in p brought about by diluting from a concentration of 18.6 to 4.1 mg/ml is approximately 0.25. Since the volume decreased 0.00046 ml/g for this change in degree of polymerization, one can calculate that the volume decrease would have been 0.0018 mg/ml if the protein had changed from a fully polymerized state (p = 1) to unpolymerized (p = 0) trimers with molecular weight 52,500. Thus, total disc-forming polymerization should result in a volume increase of approximately 0.0018 ml/g. This number must be compared with the value 0.00787 ml/g calculated in Chapter 2 for late stage polymerization. Accordingly, the volume change associated with disc-forming polymerization is less than a fourth that associated with the rod-forming process.

Structure of Double Discs

At pH values above 7.0, double discs stack to form rods, Reaction V. DURHAM *et al.*, (38, 39) reviewed the conditions which favor stacking. At $20^{\circ}C$ with a protein concentration of 5 mg/ml, stacks of two or three double discs are formed at alkaline pH values when the ionic strength is raised above about 0.6. Very much longer rods can also be obtained at alkaline *pH*, but these very long stacked double disc rods are reported by DURHAM *et al.*, to be very difficult to dissociate, once formed. Considerable work has been done on the structure of these long rods by both X-ray diffraction and electron microscopy. By applying the three-dimensional reconstruction method of DeROSIER and KLUG (40) to electron micrographs of these rods, FINCH and KLUG (41) concluded that the stacked discs are polar, that is, all protein subunits (molecular weight 17,500) are oriented the same way in the direction of the long axis of the rod - they are all top-side up. This is similar to the orientation of the subunits in the virus rod and in the typical helical rods formed under other conditions of polymerization. However, the double discs remain identifiable. FINCH and KLUG explain this as being caused by a bending or tilting downward of the outside of the subunits in the top half of a double disc, and bending or tilting upward in the bottom half. They decided against the possibility that the two halves of a double disc were oriented face-to-face for two major reasons: the angle of tilt is not the same in the two halves, and the protein subunits in both top and bottom halves of a double disc are curved in the same direction with respect to the cylinder's radii going through the centers of mass of the subunits. The crystallographers with most extensive experience with this protein have thus concluded quite definitely that in stacked double discs, the component halves are face-to-back. From this it can be inferred that this is also the structure of double discs in solution.

That inference, however, is not the logical equivalent of a direct experimental demonstration. Thermodynamics demands that the free energy decrease during the formation of double discs be substantially greater than the free energy decrease for subsequent stacking of double discs in solution. Otherwise, the double disc would not be stable. The simplest possible way to explain such thermodynamic stability is on the basis of a face-to-face orientation of the protein subunits in the two halves, a feature involved in the model proposed by the author, (42). There is no necessary inconsistency between face-to-face structure of the double disc in solution and a face-to-back structure in the dried aggregated particles studied by electron microscopy and x-ray diffraction. For the particles in solution, the dominant contribution to the negative free energy of polymerization is the entropy increase coming from the release of water molecules. This contribution must be absent in the structures viewed with the electron microscope, because the water is absent. A different kind of bonding must take over when the water content is lowered sufficiently.

This is not to say that the double disc in solution couldn't have a face-to-back structure - only that the question is still open.

Double discs can also be crystallized at *pH*=7.5 - 8.5 when the ionic strength is in the neighborhood of 0.8, (31). These crystals have a repeating unit composed of stacks of two double discs. Crystallographers interpret the structure of these crystals to indicate that the upper and lower double discs in each stack are oppositely oriented. That is, in each double disc the halves are face-to-back, but the two double discs are related to each other face-to-face.

Summary

The coat protein, TMVP, of tobacco mosaic virus has a molecular weight of 17.53×10^3. In 0.1 μ phosphate buffer at *pH* values near 6.5, the protein exists at concentrations in the neighborhood of 1 mg/ml primarily as a cyclical trimer with molecular weight of about 52.5×10^3. When the protein concentration is reduced drastically, the trimer dissociates into the basic subunit. Estimates of the standard free energy decrease for the formation of one mole of trimer from three moles of subunit cluster around 12 kcal/mole. At protein concentrations higher than 1 mg/ml, the protein exists in higher states of polymerization. The effect of protein concentration on the state of polymerization can be interpreted quite successfully in terms of Flory's mathematical development for condensation polymerization, where the cyclical trimer is the polymerizing unit. In the polymerization process discussed in this chapter, the ultimate result is the formation of double discs with 34 monomeric units. Since closed rings are formed, it is surprising that Flory's mathematical treatment can be applied, even for low degrees of polymerization. However, if it is assumed that a small amount of strain is involved in the final ring closing step, then one can account for the validity of Flory's treatment of the data.

From the effect of concentration on the state of polymerization, it is possible through the use of condensation polymerization mathematics to evaluate the equilibrium constant for the process. This equilibrium constant is the one for the reaction between two cyclical trimers to form a hexamer. It is an inherent aspect of the theory that exactly the same equilibrium constant applies to all reactions between larger particles and cyclical trimers, or between larger particles and larger particles. The extent of polymerization and, therefore, the equilibrium constant increase when the temperature is raised, demonstrating that disc-forming polymerization is an endothermic or entropy-driven process. From the variation of the equilibrium constant with temperature, one can evaluate ΔH^{o} and ΔS^{o} for the process. The van't Hoff enthalpy change thus obtained falls between 18 and 30 kcal/mole of "bond" formed, depending somewhat upon *pH*. The standard entropy changes are in the neighborhood of 100 entropy units. Direct measurement of the enthalpy change with the calorimeter yields values ranging between 25 and 30 kcal/mole of "bond" formed, values in very satisfactory agreement with the van't Hoff enthalpies.

Disc-forming polymerization is accompanied by a volume increase of 0.0018 ml/g, a value less than 1/4 of that associated with rod-forming polymerization. At most one hydrogen ion is bound during disc-forming polymerization, and the effect of *pH* on this process is very small compared to its effect on the rod-forming stage. Both disc-forming and rod-forming polymerization are favored by high ionic strength.

Chapter 4

Polymerization-Depolymerization of Tobacco Mosaic Virus Protein: Rod-Forming Polymerization

Introduction

Rod-forming polymerization of TMVP, previously designated as "high-temperature" polymerization, differs from the disc-forming stage in a number of fundamental respects. It is highly dependent upon *pH*, hydrogen ions are bound during the process, and, as pointed out in Chap. 2, the increase in volume upon polymerization is approximately 0.00787 ml/g of protein, a value more than four times that estimated for the disc-forming stage. In common with the disc-forming process, it can be interpreted in terms of Flory's mathematics for condensation polymerization. The van't Hoff enthalpies and the entropies determined for this process are strongly dependent upon *pH* and ionic strength and, under comparable conditions, are several times higher than the values for disc-forming polymerization. Rod-forming polymerization leads ultimately to the formation of rods with a helical arrangement of protein units like that in TMV. The present Chapter considers various aspects of rod-forming polymerization of TMVP, Reaction I and II of the scheme presented in Chap. 3.

Light Scatter

Much of the information obtained in the author's laboratory about rod-forming polymerization of TMVP was derived from light scatter. The actual experiments were carried out in a Beckman DU spectrophotometer using ultraviolet light at wavelengths at which TMVP does not absorb. Thus, optical density can be related solely to turbidity resulting from light scatter. Since optical density, OD, is defined as $(1/x)\ \log_{10} I_0/I$ and turbidity, τ, is defined as $(1/x)\ \ln I_0/I$, where x is the optical path length, and I_0 and I are the incident and emergent light intensities, respectively, turbidity is equal to 2.303 x optical density. In order to make certain that optical densities thus measured are indeed reasonably accurate estimates of turbidity resulting from light scatter, an unpublished control on TMV was performed by the author's colleagues, C.L. STEVENS and S. PAGLINI. Scatter was measured with a Brice-Phoenix apparatus, and was integrated over 180° to obtain the true turbidity. The turbidity of the virus sample was also calculated from the optical density as measured with the DU spectrophotometer. The two values agreed quite well.

The chamber in the spectrophometer was equipped with a jacket through which water from a thermostatically controlled bath was circulated. Thus, it was possible to measure light scatter as a function of temperature.

In the usual experiment carried out in the author's laboratory, the total concentration of TMVP was of the order of 1 mg/ml. During much of the process, the maximum dimension of the polymerized protein particles is 190 Å, a fraction of the wavelength of light used, normally 3200 Å. Thus, it is a reasonable approximation, made in the earliest analysis of our data (1), to use the simplest form of light scatter theory,

$$\tau = HcM \qquad (1)$$

Here, c is the concentration of protein in g/ml, M is the molecular weight of the protein and H is a proportionality factor depending on the refractive index of the medium, the specific refractive increment of the protein and the wavelength of the light. The appropriate value of H for TMVP and 320 mμ light is 4.23×10^{-5} $(cm/g)^2$(2).

Typical of the results obtained by the light scatter method are those shown in Fig. 1, representing the data of SMITH and LAUFFER (2). Turbidity, or optical density, is shown as a function of temperature. The data demonstrate clearly the endothermic nature of rod-forming polymerization of TMVP. The data of Fig. 1 also demonstrate that the process is reversible in a strict physico-chemical sense; measurements taken with rising temperature coincide with those taken with falling temperature.

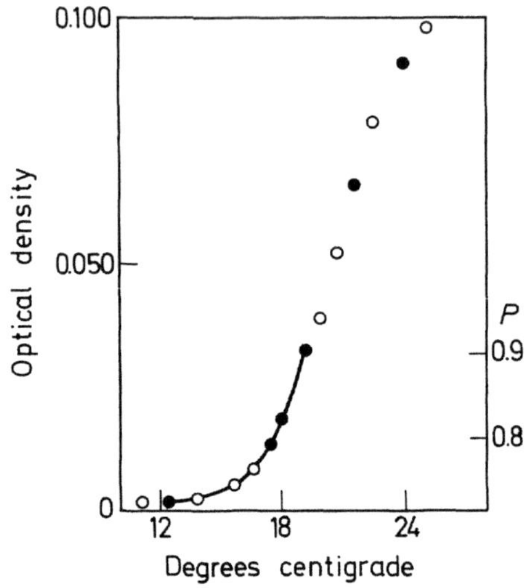

Fig. 1. Polymerization-depolymerization of 0.88 mg/ml of TMVP in phosphate buffer at *pH*=6.5 ○, rising, ●, falling temperatures. 0.1 OD ≃ 6 million $\overline{M}_w$. From LAUFFER, Max A. (1) Polymerization-Depolymerization of Tobacco Mosaic Virus Protein. *The Molecular Basis of Neoplasia* (A Collection of Papers Presented at the Fifteenth Annual Symposium on Fundamental Cancer Research, 1961, at The University of Texas M.D. Anderson Hospital and Tumor Institute at Houston, Houston, Texas). Austin: University of Texas Press, p. 192, 1962.

Several possibilities are available for interpreting such data. The least involved is to view the matter as a cooperative process somewhat akin to a phase transition taking place over a narrow temperature range. This approach has the advantage of simplicity and, as will be shown in Chap. 8, permits evaluation of thermodynamic parameters. However, it is better suited to methods which measure disappearance of small components than to a method like light scatter, which is sensitive to the size and amount of large polymers formed. An alternative procedure is to attempt to interpret the data in terms of a theory of polymerization. This also provides information concerning the van't Hoff enthalpy and the entropy of the polymerization process, but has the limitation that assumptions must be made concerning the nature of the reaction. These assumptions can be validated partially from the turbidity data themselves; obviously, for a mechanism to be correct, it is required that equations derived for it fit the data over a significant range. However, full validation of a proposed mechanism requires information beyond that inherent in the light scatter data. Since such information is difficult to obtain, this constitutes a major limitation.

Condensation Polymerization Mathematics

In any mixture containing the basic polymerizing unit with a molecular weight of M_0 and variously sized polymers of this unit the turbidity, τ_x, contributed by polymers containing x polymerizing units in the chain, is given by Eq. (2).

$$\tau_x = Hc_x M_x \tag{2}$$

Since M_x equals xM_0 and c_x/c equals xN_x/N_0, where N_x and N_0 mean the number per unit volume of x mers and of polymerizing units, respectively, and since τ_0 equals HcM_0, Eq. (3) results.

$$\tau_x/\tau_0 = x^2 N_x/N_0 \tag{3}$$

By substituting Eq. (3) of Chap. 3 and summing over all values of x, one obtains Eq. (4).

$$\tau/\tau_0 = \sum_1^\infty \tau_x/\tau_0 = (1-p)^2 \sum_1^\infty x^2 p^{(x-1)} \tag{4}$$

The summation in Eq. (4) can be evaluated. Simple division shows that $\sum_0^\infty p^x = 1/(1-p)$. Differentiation with respect to p yields $\sum_1^\infty xp^{(x-1)} = 1/(1-p)^2$. A second differentiation shows that $\sum_2^\infty x(x-1)\, p^{(x-2)} = 2/(1-p)^3$. If each successive value of x is reduced by one unit, this equation can also be written: $\sum_1^\infty (x+1)\, xp^{(x-1)} = 2/(1-p)^3 = \sum_1^\infty x^2\, p^{(x-1)} + \sum_1^\infty xp^{(x-1)}$. Therefore, $\sum_1^\infty x^2 p^{(x-1)} = 2/(1-p)^3 - 1/(1-p)^2$ or $(1+p)/(1-p)^3$. When this is substituted into Eq. (4), Eq. (5) is obtained.

$$\tau/\tau_0 = (1+p)/(1-p), \text{ or } p = (\tau/\tau_0-1)/(\tau/\tau_0+1) \tag{5}$$

When the process is reversible, as the one under consideration is, this value of p can be substituted into Eq. (8) of Chap. 3, to give Eq. (6).

$$K = \frac{(\tau/\tau_0)^2 - 1}{4[\mathrm{P}_0]} \tag{6}$$

By combining Eq. (6) with Eq. (3) of Chap. 1, one obtains Eq. (7).

$$\ln(\tau^2 - \tau_o^2) = \Delta S^o/R + \ln 4[P_o]\,\tau_o^2 - \Delta H^o/RT \qquad (7)$$

Since $(\tau - \tau_o)$ is equal to 2.303 $(OD - OD_o)$, where $(OD - OD_o)$ is the observed increment in optical density at a given temperature over the minimum value at the lowest temperature (before polymerization), Eq. (7) can be transformed into Eq. (8).

$$2 \ln 2.3\,(OD - OD_o) + \ln\,(1+2\tau_o/2.3(OD - OD_o)) = \Delta S^o/R + \ln 4[P_o]\tau_o^2 - \Delta H^o/RT \qquad (8)$$

Measurements such as those made by SMITH and LAUFFER (2) permit evaluation of $(OD - OD_o)$. However, they do not allow calculation of τ_o from OD_o because of the possibility that each sample contains a small constant amount of scattering impurity, such as denatured protein. The value of τ_o can be calculated from the protein concentration and the value of M_o, the molecular weight of the polymerizing unit, by using Eq. (1). The results of BANERJEE and LAUFFER (3) indicate that the molecular weight of TMVP in 0.1 μ phosphate buffer at pH=6.5 is 52,500. For this reason, SMITH and LAUFFER (2) assumed that a cyclical trimer with this molecular weight is the polymerizing unit. With the value of H as stated above, at a protein concentration of 0.88 mg/ml, τ_o can be calculated to have a value of 0.00195. At the same protein concentration, $[P_o]$ has a value of 1.68×10^{-5} mole/l.

Fig. 1 shows data obtained by SMITH and LAUFFER in which $OD - OD_o$ is plotted against temperature. The curve represents values of $OD - OD_o$ calculated for various temperatures according to Eq. (8), with the best fitting values of ΔH^o and ΔS^o. It is thus apparent that Eq. (8) adequately represents the beginning of rod-forming polymerization. An equation with an additional arbitrary parameter, OD_m, the maximum optical density, is required to fit the data over the entire range of temperature (2). This is presented as Eq. (9).

$$2 \ln 2.3(OD - OD_o) + \ln(1 + 2\tau_o/2.3(OD - OD_o)) - 2 \ln 2.3(OD_m - OD) = \Delta S^o/R + \ln 4[P_o]\tau_o^2 - 2 \ln(\tau_o + 2.3(OD_m - OD_o)) - \Delta H^o/RT \qquad (9)$$

Eq. (9) differs from Eq. (8) in having an additional variable on the left and an additional constant on the right, both involving the new arbitrary parameter, OD_m. While this equation was originally developed on theoretical grounds, it was later found that a number of different theories led to indistinguishable final results. Therefore, Eq. (9) must be considered empirical. Data can be fitted with the computer to Eq. (9) by an iteration process. With assigned values of τ_0 and $[P_0]$, the values of OD_m and of ΔH^0 and ΔS^0 which give the best fit can be found, and from these, ΔF^0 at a particular temperature can be calculated.

The values of ΔH^0, ΔS^0 and ΔF^0 depend to a small extent upon the assumed value of the molecular weight of the minimum polymerizing unit. This is true because both $[P_0]$ and τ_0 are calculated from the protein concentration and the assumed value of M_0. This potential source of error has been discussed by LAUFFER (4). It was shown that if only data for which $2.3(OD - OD_0)$ is greater than $4\tau_0$ are used, the value of ΔH^0 is nearly independent of the assumed value of τ_0. In this case, the calculated values of ΔS^0 and ΔF^0 will be linearly dependent upon the assumed value of $\ln \tau_0$. A 5-fold change in the assumed value of τ_0 would result in a change in the calculated value of ΔF^0 of 1 kcal/mole.

It is evident, according to Eq. (8) and (9), that it should be possible to evaluate ΔH^0 and ΔS^0 from measurements of $(OD - OD_0)$ at different temperatures. If one plots the variables on the right of Eq. (9) against $1/T$, a straight line should result. Such a plot is represented in Fig. 2, which shows the fit of all of the data points found in Fig. 1. If, instead, one plotted the right hand variables from Eq. (8) against $1/T$, the data at the lower temperatures would fit the same graph, but those at higher temperatures would depart significantly from it. Thus, essentially the same values of ΔH^0 and ΔS^0 can be obtainet, whether one uses the theoretical Eq. (8) or empirical Eq. (9). The advantage of Eq. (9) is that it permits use of all of the data, not only those obtained, at low degrees of polymerization. Another way of saying this is that Eq. (9), with the properly assigned arbitrary parameters, is an adequate way of describing all of the data obtained in a given experiment. Thus, to describe the results of an experiment, one has the option of drawing a graph or of listing the arbitrary parameters found in Eq. (9).

It is quite apparent that data such as those obtained by SMITH and LAUFFER for polymerization of TMVP in 0.1 μ phosphate buffer at pH=6.5 fit the simple theory based on condensation polymerization for rod-forming polymerization. The simple theory without any empirical corrections fits the data in the temperature range 12-19^{0}C and up to a value of $OD - OD_0$ of 0.03. This theory is that units, assumed by SMITH and LAUFFER to be cyclical trimers with a molec-

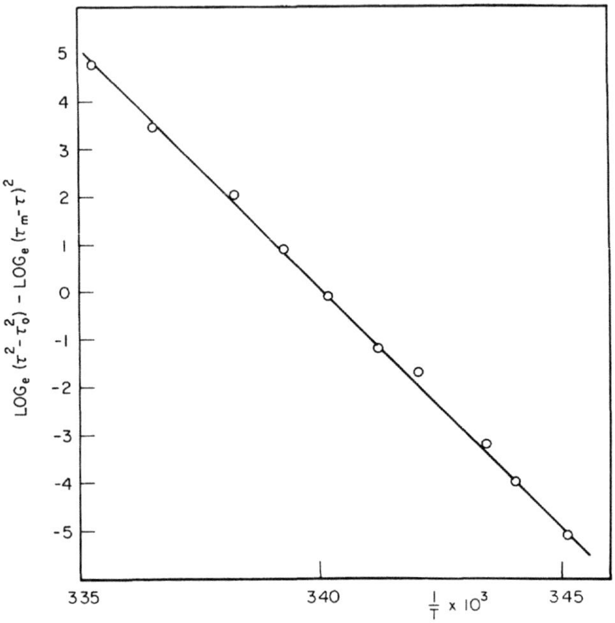

Fig. 2. Evaluation of τ_m, ΔH and ΔS^{o} for polymerization of TMVP. Fit of data of Fig.1 to Eq. (9). From SMITH and LAUFFER (2). Reprinted by permission of the copyright owner, The American Chemical Society

ular weight of 52,500, are polymerizing linearly (not necessarily in a straight line) in a process mathematically akin to condensation polymerization of simpler molecules. This assumption corresponds more or less to Reaction I in the scheme presented in Chap. 3. The values of ΔH^{o} and ΔS^{o} corresponding to this particular fit are 187 kcal/mole and 708 entropy units, respectively. Both are positive, as deduced in 1958 from the first observations of LAUFFER *et. al.* (5) on TMVP polymerization.

The reason why SMITH and LAUFFER (2) assumed that the cyclical trimer is the polymerizing unit has already been presented; at *pH*=6.5 in 0.1 μ phosphate buffer at temperatures in the neighborhood of 10°C, this is the predominant particle present. However, SMITH and LAUFFER were aware of the fact, published by our own laboratory in 1958 (5), that at 12°C TMVP in 0.1 μ phosphate buffer at *pH*=6.5 exhibited a small sedimenting boundary in the ultracentrifuge with a sedimentation coefficient of approximately 25 *S*, as well as a considerably larger boundary sedimenting at about 4 *S*. EISKAMP (6) reported similar findings. This faster boundary can be identified with particles of approximately the size of double discs. SMITH and LAUFFER, therefore, recognized the possibility that the actual polymerization process being observed by light scatter might be the assembly of larger units, perhaps units as large as the double disc. They pointed out that had their data been interpreted on this assumption, no major difference in parameters, ΔH^{o} and ΔS^{o}, would have resulted. CASPAR (7) did interpret the data of SMITH and LAUFFER on the assumption that the double disc is an intermediate in the polymerization process.

Several facts have accumulated during recent years to suggest that light scatter measurements might indeed depend upon polymerization of double discs or structures of comparable size, Reaction II of the scheme presented in Chap.3. PAGLINI and LAUFFER (8) confirmed and extended the results of BANERJEE and LAUFFER (3). On the one hand, in 0.1 ionic strength phosphate and barbital buffers at *pH* values of 6.5 and 7.5, polymerization brought about by changing concentration fitted very well the assumption that trimers with molecular weight of 52,500 are the polymerizing unit. Most of the data were obtained at 4°C, but a very good fit was also obtained at 11.7^{o}C in phosphate buffer at *pH*=6.5.

Data obtained in 0.1 ionic strength phosphate buffer at *pH*=7.5 at 20 and 25° could be forced to fit this hypothesis, but the fit was not very good. On the other hand, data obtained in 0.1 ionic strength pyrophosphate buffer at *pH*=7.3 and in 0.1 ionic strength phosphate buffer at *pH*=6.0 did not fit this hypothesis. Neither did data obtained with D_2O as solvent. All of the experiments which did not fit the earlier theory indicated a higher degree of polymerization than the cyclical trimer for the starting material, even at very low concentrations.

DURHAM, *et. al*, (9, 10) reported a *pH*-drop experiment in which polymerization was followed with the electron microscope. A sample of TMVP, which at a concentration of 5 mg/ml at *pH*=7.0 showed primarily a 20 S component in the ultracentrifuge, exhibited mainly double discs in the electron microscope. When this solution was diluted rapidly in acetate buffer at *pH*=5, polymerization resulted. A sample taken five seconds after dilution and subjected to electron microscopy showed short rods with spacings or defects corresponding to each two turns of the helix. When the samples were taken for electron microscopy 15 minutes after dilution, the rods were longer and there were fewer defects. After 18 hours, almost perfect helical rods were obtained. DURHAM *et. al.* interpreted their results on the assumption that double discs are converted into "lock washers" of the same molecular weight when the *pH* is lowered. These "lock washers" then polymerize to form imperfect helices because of poor fit, which later anneal into perfect helicies. However, no direct demonstration of the existence of "lock washers" was claimed. The significant aspect of this experiment is that rod-forming polymerization is shown, under the conditions used, to be the polymerization of particles of the same size as double discs.

An even more plausible model than the "lock washer", which has a break in the central strand of the two-turn helix, is a two-turn spiral. There is no reason why the two ends at the break in the central strand of the "lock washer" should not be joined; they are close together and the thermodynamic factors favoring such joining are identical with those favoring every other lateral union in the two-turn structure. This is the model described previously by the author (4) as a short, closed helix or cylinder or short rod with ratio of length to diameter of approximately one half. For convenience, it will be called a double spiral. Like "lock washers", double spirals are hypothetical; no direct experimental evidence for the existence of such structures as stable intermediates is available. The *pH*-drop experiment corresponds to Reaction IV followed by II in the scheme.

Evidence pointing in the same direction was also obtained by WESTOVER (11). He found in sedimentation equilibrium studies on TMVP in the early stages of polymerization that, regardless of the concentration of protein introduced into the ultracentrifuge cell, at equilibrium there was a definite maximum "cut-off" concentration remaining. He assumed that this resulted from the formation of a polymer with a sedimentation coefficient like that of double discs which was removed from the solution by sedimentation. He assumed further that this "insoluble" polymer was formed at some definite value of p, the parameter in Eq. (8) of Chap. 3 defined as the probability that a site on a polymerizing unit is combined. On the basis of these assumptions, the equilibrium constant for the polymerization of the small units to form the insoluble" polymer is, according to Eq. (8) of Chap. 3, proportional to the reciprocal of the "cut-off" concentration. When this result is substituted into Eq. (3) of Chap. 1, one obtains,

$$-\ln C_c = -\Delta F^{\circ}/RT - \ln \alpha = \Delta S^{\circ}/R - \ln \alpha - (\Delta H^{\circ}/R)(1/T) \qquad (10)$$

Here C_c means the "cut-off" concentration and α is the proportionality constant. This equation should be compared with Eq. (2) of Chap. 8 and Eq. (6)

of Chap. 10, equations of the same general form derived on different assumptions. When Eq. (10) and others like it are differentiated with respect to T, one obtains a version of the Clapeyron-Clausius equation, known since 1834. By plotting $- \ln C_e$ against $1/T$, WESTOVER showed that ΔH^0 for the polymerization of small units was of the order of magnitude of 30 - 40 kcal/mole for polymerization in 0.1 μ phosphate buffer at pH=6.5 and also for polymerization in 0.1 M pyrophosphate buffer at pH=7.8. He also obtained limited data for polymerization of TMVP in 0.1 μ phosphate buffer at pH=6.0, which indicate a value of ΔH of the same order of magnitude. Furthermore, TANIGUCHI (12) showed by sedimentation velocity that at low protein concentration only a 4 S boundary is present, but when the loading concentration is increased, a critical concentration is reached at which 10 S to 20 S boundaries appear. This critical concentration is lowered by increasing temperature. A value of 77 kcal/mole for the polymerization of 4 S material was estimated from these results. Since the studies here described of necessity involve the polymerization of small units, the finding of a ΔH^0 value of the order of 30 to 77 kcal/mole is evidence in favor of the assumption that the polymerization process studied by light scatter, which yields higher values of ΔH^0, must apply to some other reaction.

As will be described in the next chapter, the polymerization pattern of the Dahlemense strain of TMV protein, except for differences with respect to pH, greatly resembles that of TMVP (3). Even at pH=6.5, the molecular weight of the material before temperature-induced polymerization was as high as or higher than that of the double disc. No evidence was obtained to indicate that particles smaller than double discs are involved in rod-forming polymerization of this material. It seems unreasonable to postulate one mechanism for polymerization of Dahlemense virus protein and another for the polymerization of TMVP. Furthermore, it was found that at pH=6.0, the starting material for temperatur-induced polymerization of TMVP had a molecular weight corresponding to double discs or somewhat larger particles (13).

For all these reasons, it seems probable that the actual process followed by light scatter in rod-forming polymerization of TMVP involves the polymerization of double spirals of the same molecular weight as double discs, Reaction II. The author had been reluctant to adopt this point of view, because the results obtained at pH=6.5 by SMITH and LAUFFER (2) and BANERJEE and LAUFFER (3) clearly indicate that the starting material had a molecular weight of 52,500 and because the mathematics of condensation polymerization applied accurately over a molecular weight range from about 50,000 to 1 million. It is necessary to explain how this can be the case if the light scatter experiments used to follow the polymerization are actually sensitive to the polymerization of double spirals with a molecular weight of 596,000. A possibility is that double spirals first form from the low molecular weight material and then immediately polymerize further, even before most of the low molecular weight material is used up. STEVENS (14) has shown that this kind of assumption can fit the facts.

STEVENS assumed three steps in the polymerization of TMVP in 0.1 μ phosphate buffer at *pH*=6.5. The first step represents the polymerization of cyclical trimers in accordance with the assumptions appropriate for condensation polymerization. The second step represents the addition of one cyclical trimer to a curved chain to close the double disc or form the double spiral. The third step represents the polymerization of these double spirals in accordance with the assumptions appropriate for Flory's treatment of condensation polymerization. Each step is assumed to have its own equilibrium constant, and these equilibrium constants are considered to be functions of temperature. STEVENS' mathematical treatment is an extension of the treatment used by the author (4) to explain why Flory's mathematics can apply to the disc-forming polymerization. STEVENS' final equation for turbidity as a function of concentration involves three equilibrium constants and their variation with temperature plus an unmeasurable parameter, the transient concentration of cyclical trimer. However, he was able to show with the aid of the computer that these assumptions could lead to the result that polymerization of double spirals can begin even when a substantial fraction of the material is in a much lower state of polymerization. Furthermore, he was able, by the proper selection of arbitrary constants, to obtain theoretical curves in reasonable agreement with the experimental results of SMITH and LAUFFER. This theoretical treatment does explain how the results obtained by SMITH and LAUFFER for TMVP in 0.1 μ phosphate buffer at *pH*=6.5 can fit Flory's mathematics for condensation polymerization of units with molecular weight 52,500, even if the turbidity measurements are actually primarily dependent on the polymerization of aggregates with a molecular weight of 596,000. However, one of the fruits of Stevens' theoretical treatment is that the parameters ΔH^{o} and ΔS^{o} evaluated by fitting the data to Eq. (13) no longer correspond to unique chemical reactions.

The finding of WESTOVER (11), however, provides an alternative and even elegant method for analyzing the result of light scatter experiments. Increments in turbidity, 2.3 ($OD - OD_0$), of Eqs. (8) and (9) are attributed to polymerization of particles with molecular weight 596,000. $[P_0]$ and τ_0, however, are considered as variables which have different values at different temperatures. The values at each temperature can be determined from the "cut-off" concentrations of WESTOVER. The concentration of material contributing to the increment in turbidity is the initial concentration of protein in the solution minus the "cut-off" concentration at the particular temperature under consideration. $[P_0]$ is this concentration expressed as grams per liter divided by the molecular weight of the polymerizing unit to which light scatter is sensitive, 596,000. τ_0 is equal to this concentration multiplied by H and the same molecular weight. While the arithmetic is tedious, such calculations have been made on the data from two light scatter experiments, both carried out in 0.1 μ phosphate buffer, one at pH=6.0 and the other at pH=6.5. Data treated this way fitted Eqs. (8) and (9) just as well as the same data treated on the assumption that $[P_0]$ and τ_0 are constants calculated from the total protein concentration and a molecular weight of 596,000. Furthermore, the values for ΔH^0 obtained by the more complicated treatment were indistinguishable from those obtained by the simpler treatment. This finding shows that if the data are treated by the simpler method on the assumption that M_0 is 596,000, reliable thermodynamic parameters will be obtained for the further polymerization of double spirals.

ΔH^0 and ΔS^0 were originally calculated from the light scatter data published in References (2) and (3) and in other references to be cited later in this chapter on the assumption that M_0 is 52,500. Greatly different values are not obtained when the data are recalculated on the assumption that M_0 is 596,000. For the purpose of the present chapter, the original calculated values are retained, but they are now interpreted to be approximations rather than exact values of ΔH^0 and ΔS^0 for Reaction II, the polymerization of double spirals, particles of the same molecular weights as double discs.

Cooperativity

The polymerization of TMVP has been described as a cooperative process (7). The reasons become obvious when it is considered that the transition from a solution of protein subunits in their monomeric form to a suspension of long rods with very high molecular weight resembles greatly the crystallization of a protein. Such phase transitions are typical representatives of cooperative processes. In a crystal of a molecular substance, one molecule is surrounded by a number of neighbors and is bonded in some way to each. In the rod-like TMVP polymer, a single protein subunit is surrounded by six neighbors and is presumably "bonded" to each. Multiple bonding is a characteristic of cooperative processes. If all "bonds" between subunits are equivalent, the free energy decrease when two monomeric subunits come together must be only about half as great as when a third subunit adds to the dimer to form cyclical trimer, because in the first case there is only one point of contact or one "bond", while in the second, there are two points of contact or two "bonds" for each monomeric unit added. Again, this is typical of what one finds in cooperative processes; the reactivity of a unit is enhanced when it is already combined. Another typical aspect of cooperative processes is that the equilibrium is shifted from a highly dispersed to a highly condensed state over a narrow temperature range. The data shown in Fig.1 demonstrate that rod-forming polymerization of TMVP takes place over a range of only a few degrees C. Furthermore, it is clear that the formation of each contact between protein subunits involves the simultaneous release of a large number of water molecules. Theses water molecules behave cooperatively. Finally, intermediates like cyclical trimers or double spirals seem to polymerize as units. If one focuses attention on the subunits making up the intermediates, one can then say that they polymerize cooperatively because they all react together as a unit.

In spite of all of this, the description of TMVP polymerization as a cooperative process obscures some aspects of reality while focusing attention on others. Flory's mathematics can't possibly apply to a process which is cooperative in every sense in which the term is used, because Flory's equations apply only to the situation in which the reactivity of a unit is independent of whether it exists free in solution or at the end of a polymer chain. Flory's mathematical formulation does seem to apply over a wide range of weight average molecular weight, both in disc-forming and in rod-forming polymerization. Thus, there are steps in both of these polymerization processes for which it is definitely misleading to emphasize cooperativity without at least specifying that the cooperativity is at a low level of organization, rather than at the level of size of the polymerizing units themselves. The large temperature coefficient illustrated by the results shown in Fig. 1 can be explained by the fact that large numbers of water molecules are released whenever a union between partially polymerized particles occurs. This provides a high positive enthalpy, which, in itself, accounts for the major change in state of polymerization over a narrow temperature range. It is thus unnecessary that there be any cooperativity at all at the polymer level to account for this steep transition. Obviously, however, cooperativity at the small molecule level is certainly involved.

Because of the fact that describing TMVP polymerization as cooperative can be misleading as well as instructive, the term will be used sparingly in the present work. This word has too many connotations, some of which are not applicable. It is hoped that others who use the term frequently can be encouraged to define precisely what is meant by it in the context in which it is being used.

Effect of pH

It has already been pointed out in Chap. 3 that changes in *pH* have a small effect only on disc-forming polymerization. In contrast, *pH* has an enormous effect on rod-forming polymerization. The first evidence of this was obtained in 1958 (5), when it was shown that at *pH*=5.0, TMVP was highly polymerized at all temperatures studied; at *pH*=6.5, it was polymerized above and depolymerized below about 11°C; at *pH*=7.7, it was depolymerized at all temperatures studied. ANSEVIN and LAUFFER (15) found that endothermic polymerization could also be observed at *pH* values below the isoelectric point of TMVP. Under these conditions, the reaction is complicated by denaturation and other side-effects.

KHALIL and LAUFFER (16) studied the effect on TMVP polymerization of ionic strength and *pH* and also of ionic strength and *pD* for solutions containing D_2O instead of H_2O. The light scatter method of SMITH and LAUFFER (2) was used. The data were fitted to Eq. (9) with the assumption that the polymerizing unit has a molecular weight of 52,500; in this manner ΔH°, ΔS° and τ_m were evaluated. It was found that ΔH° and ΔS° varied somewhat erratically with *pH* or *pD*. However, the data obtained by KAHLIL and LAUFFER were sufficiently extensive to establish definite trends. On the average, increase in *pH* or *pD* resulted in reductions both in ΔH° and ΔS°.

In contrast with ΔH° and ΔS°, ΔF° varies more regularly. This is not surprising, since ΔF° is directly related to the equilibrium constant, *K*. If, for a given value of ΔF^o, an error is made in evaluating ΔH^o, a compensating error will result in the calculated value of ΔS^o. For convenience in calculating, a temperature of 290° K was chosen. Even though the data still exhibit considerable scatter, the trend is clear and is approximately independent of ionic strength. Accordingly, $\Delta F^o{}_{290}$ values at all ionic strengths between 0.05 and 0.16 calculated from the data of KHALIL and LAUFFER were averaged at each *pH* value between 6.0 and 6.6. The results are shown in Fig. 3, where $-\Delta F^o{}_{290}$ is plotted against *pH*. For one mole of "bond", $d\Delta F^o/dpH$ has an approximate value of +7.2 kcal per *pH* unit in the *pH* range 6.0 - 6.6. Similar treatment of data obtained with D_2O as solvent showed that $d\Delta F^o/dpD$ has a value of +15.5 for the range *pD* 6.4 - 7.0. Had the thermodynamic parameters been evaluated on the assumption that the polymerizing unit has a molecular weight of 596,000, each value calculated would be slightly different, but the trends would be the same.

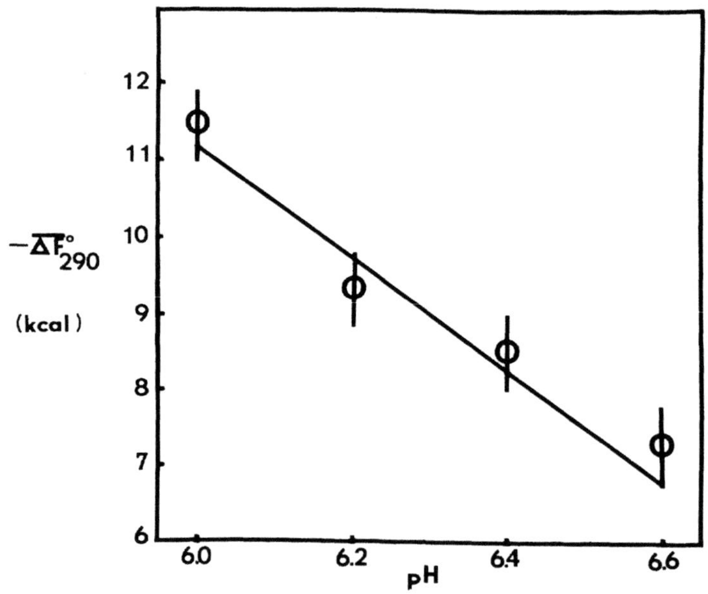

Fig. 3. Mean value of the standard free energy at 290^oK for polymerization of TMVP plotted as a function of *pH*. Data of KHALIL and LAUFFER (16)

In contrast with the somewhat erratic fluctuations in ΔH^o and ΔS^o, the temperature range over which polymerization takes place was found to vary quite regularly with *pH* (16). For this reason, a characteristic temperature, T^*, was chosen for each polymerization. The characteristic temperature is the temperature at some characteristic value of $(OD - OD_o)$. SHALABY and LAUFFER (17) defined T^* as the temperature at that value of $(OD - OD_o)$ which corresponds to a turbidity equal to 0.1 τ_m divided by 0.1 + τ_m. It must be admitted that this is entirely arbitrary. However, the choice has two virtues. First, it can be shown by substituting this value into Eq. (9) that, for a TMVP concentration of 1 mg/ml, T^* is the temperature corresponding to a constant value of ln K equal to 16.7. This assertion can be verified by substituting τ for 2.3 OD, preserving the subscripts, and ln K for $\Delta S^o/R - \Delta H^o/RT$ into Eq. (9), and then simplifying by considering that $(\tau^2 - \tau_o^2)$ is approximately τ^2. The second advantage of this arbitrary assignment is that the value of $(OD - OD_o)$ thereby chosen is high enough to fall in the range where the data are reliable. When characteristic temperatures thus defined are plotted against pH or pD for ionic strength values ranging between 0.05 and 0.16, the results are as shown in Fig. 4, reproduced from LAUFFER and STEVENS (18). Clearly, at all values of ionic strength, polymerization takes place at lower temperatures, the lower the pH or pD.

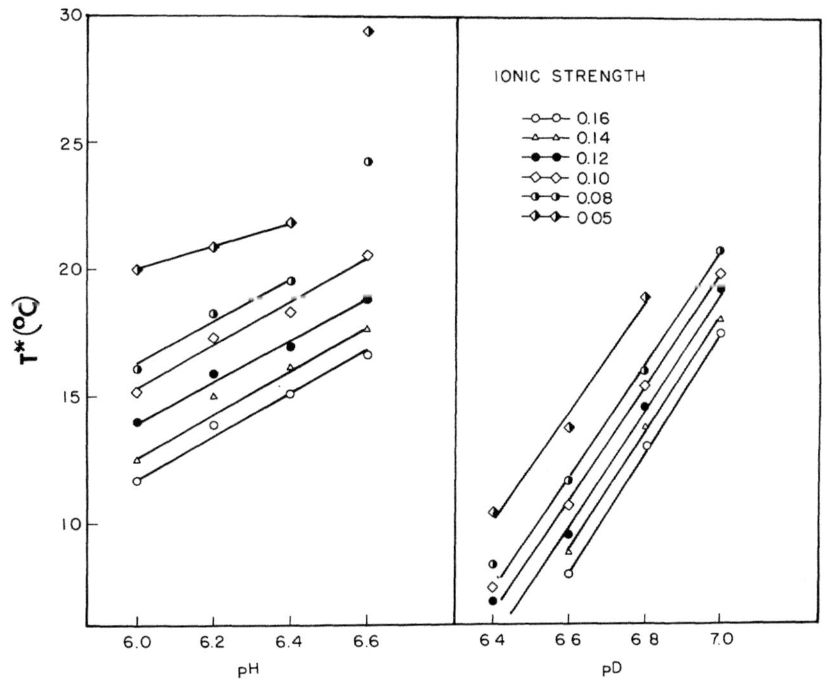

Fig. 4. Characteristic temperature (T^*) in °C plotted *vs.* pH or pD for polymerization of TMVP. Data of KHALIL and LAUFFER (16). Reprinted from LAUFFER and STEVENS (18). Reprinted by permission of the copyright owner, Academic Press, Inc.

The values of τ_m, obtained by the curve fitting technique described earlier, showed a consistent drift with *pH* or *pD*, but no consistent trend for variations in ionic strength. In Table 1 are found the average values of τ_m at all values of ionic strength shown in relation to *pH* and *pD*. The results demonstrate an almost completely consistent reduction in τ_m as *pH* or *pD* are increased. Except at 6.4, the values of τ_m are higher for a particular value of *pD* than for the corresponding values of *pH*.

Table 1. Values of τ_m at various values of *pH* and *pD*

pH	τ_m	*pD*	τ_m
6.0	.250	---	----
6.2	.228	---	----
6.4	.215	6.4	.208
6.6	.137	6.6	.215
6.8	.076	6.8	.172
---	----	7.0	.114

When TMVP polymerizes in an unbuffered medium as a result of an increase in temperature, there is a sharp increase in *pH*. This result is demonstrated in Fig. 5, published originally by ANSEVIN, STEVENS and LAUFFER (19). This increase in *pH* is the result of the binding of hydrogen ions by the protein when it polymerizes, a fact demonstrated by the acid-base titration data of SCHEELE and LAUFFER (20), shown in Fig. 6. Here it can be seen that there is a distinct difference between the titration of relatively unpolymerized TMVP at 4°C and polymerized TMVP at 20°C in the *pH* region 6-7. The number of hydrogen ions bound per monomeric protein unit as a result of polymerization at

Fig. 6. Acid-base titration curves of TMVP at 4° and 20°C. HCl (□) at 4° and (○) 20°C; KOH (△) at 4° and (○) 20°C. From SCHEELE and LAUFFER (20). Reprinted by permission of the copyright owner, The American Chemical Society ▶

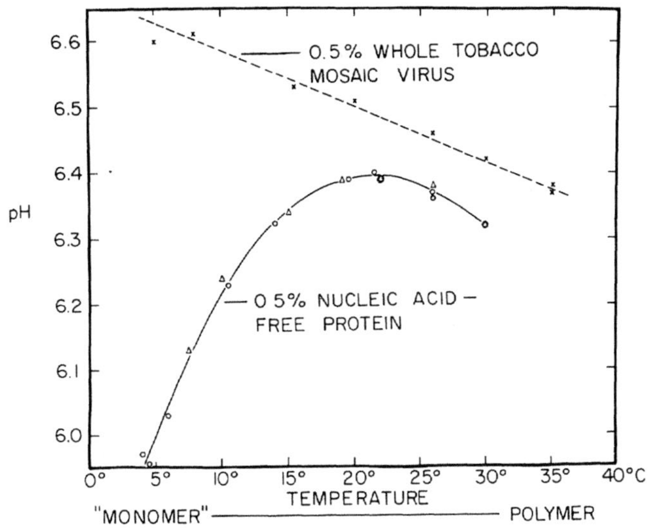

Fig. 5. *pH*-meter reading of unbuffered TMVP and unbuffered virus in 0.1 μ NaCl plotted as a function of temperature and extent of polymerization (monomer is considered to be the cyclical trimer of the protein subunit and polymer is protein rods). Circles represent descending temperatures and triangles ascending temperatures for data with TMVP; crosses represent TMVP control. From ANSEVIN, STEVENS and LAUFFER (19). Reprinted by permission of the copyright owner, The American Chemical Society

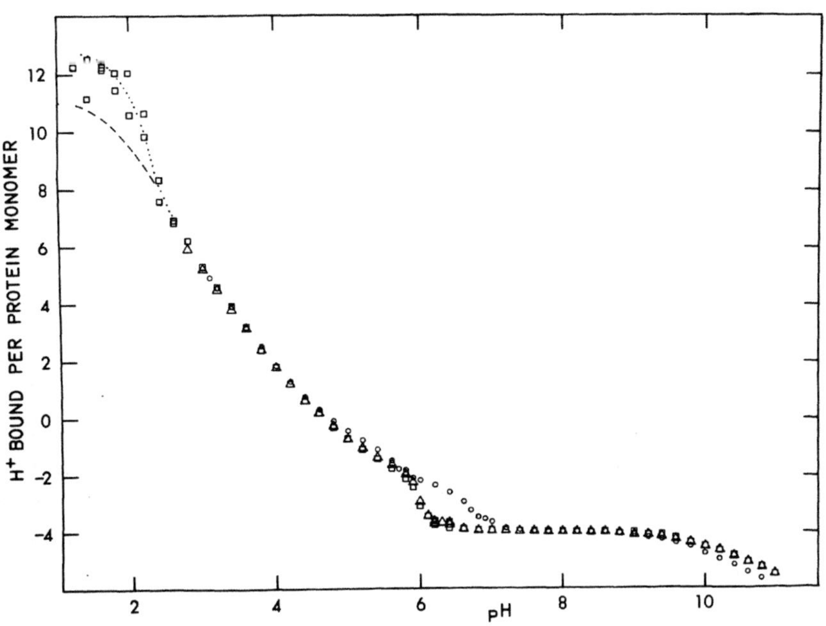

any particular *pH* value can be determined from the difference between the two curves in this region. At *pH*=6.5, about one hydrogen ion is bound per chemical subunit when polymerization is brought about by raising the temperature from $4^{o}C$ to $20^{o}C$.

When a typical carboxyl group on a protein ionizes, ΔH^{o} is 1110 cal/mole and ΔS^{o}, largely attributable to binding of water by electrostriction, is -17.9 e.u. When, however, the protein ionizes in a buffer at constant *pH*, there is a simultaneous reaction between buffer ion and hydrogen ion to produce buffer acid (19). The net ΔH^{o} and ΔS^{o} in phosphate buffer are +123 cal/mole and +11.72 e.u. when buffer salt, buffer acid, ionized and nonionized carboxyl groups are all at unit activity. When the ionized and nonionized carboxyl groups are at unit activity, but the buffer is at some specific *pH* value in which the buffer acid and buffer salt are not at the same activity[1], different values will be obtained. In phosphate buffer at *pH*=6.5, ΔS^{o} is about 8.35

---

[1] The state here described illustrates a problem encountered by those who study complicated systems like polymerizing proteins, a problem not shared by those physical chemists who work with simple, well-defined reactions in which the stoichiometric amounts of all participants are known. For such simple systems, the standard state is defined as that at which every participant is at unit activity. As a corollary, equilibrium constants are evaluated from the concentrations in equilibrium systems of all participants. For the more complicated systems, one usually does not have sufficient information to write a complete stoichiometric chemical equation. The polymerization of TMVP is a good example. Equilibrium constants are calculated from changes in the amounts of polymerizing material and of polymer, ignoring in the calculation all other possible reactants such as hydrogen ions, but keeping known variables constant experimentally. There are two consequences. One is that the equilibrium constants evaluated may differ at different *pH* values, at different ionic strengths, etc. It is possible to get around this difficulty by specifying the *pH*, the ionic strength and any other solution variable known to affect the equilibrium constant. The other consequence is that the standard state refers only to unit activity of polymerizing material and of all polymer species. Unlike the case with simple reactions of completely known stoichiometry, this standard state is incomplete; *pH*, ionic strength and other known variables must be specified.

entropy units per mole of hydrogen ion dissociated and ΔH is still 123 cal/mole. ΔS° increase 4.56 e.u. per unit increase in *pH*. The free energy change at 290° K is thus -2300 cal/mole. For the binding of a mole hydrogen ion by TMVP in a phosphate buffer at *pH*=6.5, ΔF°_{290} is +2300 cal. Since the net free energy change for polymerization must be negative, the reaction being spontaneous, it is clear that the binding of hydrogen ions opposes polymerization. Contrary to the view expressed by DURHAM *et al.* (9), the polymerization process that occurs in the neighborhood of *pH*=6.5 cannot be driven by adding protons. The addition of protons involves an increase in free energy. Some other source of free energy must be available to permit the protein to capture protons.

This point is so important that it deserves elaboration. First, it must be observed that, since in any buffer system the hydrogen ion activity is at equilibrium with the salt and the acid, the actual free energy of dissociating or binding a hydrogen ion by the buffer system is zero. Thus, the standard free energy, defined as above, for the binding of one mole of hydrogen ion by TMVP is +2300 cal, regardless of the buffer system and even in an unbuffered system, provided only that *pH* is 6.5. The second point to be observed is that whatever bind hydrogen ions during polymerization of TMVP, they have *pK* values well below 6.5 when the protein is in the relatively unpolymerized state. The titration data of SCHEELE and LAUFFER (20) obtained at 4°C show this clearly. It is usually assumed that side-chain carboxyl groups on protein have *pK* values in the neighborhood of 4.7. When the values for ΔH° and ΔS° for protein carboxyl ionization mentioned above are substituted into Eq. (1) of Chap. 1, ΔF°_{290} can be calculated to be +6290 cal/mole. With the aid of Eq. (2) of Chap. 1, $-\log_{10} K_{290}$, or *pK*, can be shown to be 4.76. As was pointed out above, when a carboxyl group with such a *pK* value binds a hydrogen ion at *pH*=6.5, the increase in free energy is +2300 cal/mole.

In the absence of specific kinetic data, one can assume many possible paths which might be taken by relatively unpolymerized protein with ionized carboxyl groups with *pK*=4.76 to become polymerized protein in a particular configuration, such as the double spirals, with a hydrogen ion bound by each carboxyl group. The overall free energy change, ΔF°, is the same for all possible equilibrium paths. One such path is for relatively unpolymerized protein with *pK* of 4.76 to polymerize to form a structure (such as a double disc) with ionized carboxyl groups with *pK*=4.76, followed by binding of hydrogen ions with an increase of free energy of 2300 cal/mole to permit a shift in conformation (to the double spiral state). The binding of the hydrogen ion must be coupled with a change involving a sufficiently high negative free energy change to overcome it. A second possibility is the polymerization of protein with ionized carboxyl groups with pK=4.76 to a structure (double discs) with pK=4.76, followed by an intermediate structural change which raises the pK of the carboxyl groups to, let us say, 7, followed by the binding of hydrogen ions at pH=6.5 and transition to a different polymerized form (double spiral). In this case, the change in structure which brings about the change in pK of the carboxyl groups involves an increase ($-RT\ \Delta\ \ln K$) in ΔF° of 2960 cal/mole. The binding of hydrogen ions by such groups at pH=6.5 would involve a decrease in ΔF° of 660 cal/mole. The net for the change in pK and the binding of the hydrogen ions is +2300 cal/mole, exactly as before. Here again, the process involving the increase in pK must be coupled with a process having a sufficiently large negative free energy change. One can imagine alternative reversible pathways endlessly. All of them have the same feature in common; each must be coupled with some process involving a sufficiently large decrease in free energy to over-balance the net positive free energy of 2300 cal/mole of hydrogen ion bound. The necessity for hydrogen ions to be bound is a thermodynamic liability in the conversion of relatively unpolymerized TMVP to typical helically polymerized protein at all pH values above the pK of carboxyl groups present on the protein in the relatively unpolymerized state.

In phosphate buffer at pH=6.0, ΔH°, ΔS° and ΔF°_{290} are -123, -6.10 and +1647 respectively, for the binding of one mole of hydrogen ion. Hence, the opposition to polymerization caused by hydrogen ion binding drops from 2.300 kcal/mole to 1.647 kcal/mole of hydrogen ion bound when the pH is dropped from 6.5 to 6.0. This corresponds to a value for $d\Delta F^{\circ}/dpH$ of 1.31 kcal/mole of H^+ bound/pH unit. Since each double spiral is composed of 34 protein subunits, and since approximately one H^+ ion is bound per subunit during polymerization at pH=6.5, this effect is more than large enough to account for the change in ΔF° with change in pH shown in Fig. 3. In fact, if all H^+ were bound in Reaction II, $d\Delta F^{\circ}/dpH$ should have a value +44.5 kcal/mole per pH unit. The actual value derived from Fig. 3 is +7.2. This would indicate that only a fraction of the H^+ ion are bound in the step, the rest being bound is Reaction I. It is interesting to observe that this is purely an entropic effect and is related entirely to the different entropy changes at different pH values experienced by the buffer in restoring to the solution the hydrogen ions bound by the protein.

There remains to be explained the strong dependence shown in Table 1 of τ_m upon *pH* or *pD*. Whatever the uncertainties in the estimation of τ_m, its value must be at least some sort of crude estimate of the maximum size that particles achieve at various values of *pH* or *pD*. In general, larger particles are ultimately formed at lower values of *pH* or *pD*. The titration results of SCHEELE and LAUFFER (20) show that polymerized TMVP is combined with fewer ions at *pH*=6.0 than at *pH*=6.8. Presumably, therefore, the net negative charge on a rod of a given length is greater at *pH*=6.8 than at *pH*=6.0. Thus, there is a strong inverse correlation between net negative charge and maximum degree of polymerization. This fact suggests that the explanation might be found in terms of the contribution of electrostatic repulsion to the free energy of polymerization of shorts rods into longer rods. This problem was considered in detail by the author (4). In order to estimate the electrostatic contribution to the free energy of the reaction between two short rods of equal length to form a rod of double the original length, the author derived an equation for the electrical work, W_{el} for charging a cylinder. The model chosen was a cylinder of length, L, and radius, R, with a uniform charge, Q, on its curved surface. This model differs from a charged cylindrical conductor, where the charges are free to move and the charge per unit length will therefore be greater at the ends than in the middle, because there is less mutual repulsion in charges at the end. The original publication of this theory contains obvious errors in type-setting; the correct form is shown in Eq. (11).

$$W_{el} = \frac{Q^2 \ln (1+1/\kappa a)}{DL^2} \quad (R + 1/2\kappa)\left[\sqrt{1+(L/(R+1/2\kappa))^2} - 1\right] \qquad (11)$$

D means the dielectric constant, κ is the Debye-Hückel constant and a is the radius of closest approach, i.e., the sum of the radius of the cylinder and the mean radius of the counter ions. This equation differs somewhat from previous equations derived by GORIN (21) and by HILL (22). All three equations become identical when the length of the cylinder is enormously greater than its radius.

A slightly hydrated, polymerized TMVP rod can be assumed to have a value of R of 9.3×10^{-7} cm and a value of a equal to 9.5×10^{-7} cm. There are 16 1/3 protein subunits per 23 Å or 0.71 subunits per 10^{-8} cm; therefore Z/L is $0.71\, z \times 10^8$ ionic charges/cm, where z is the charge on a protein subunit and Z is the total number of ionic charges. The total charge, Q, is $Z\,\varepsilon$, where ε is the electronic charge, 4.8×10^{-10} absolute electrostatic units. At 25°C and

0.1 μ, κ has a value of 1.04×10^7 cm^{-1} and *D* has a value of 78.54. When one substitutes these values into Eq. (11) and converts from particles to moles by multiplying by Avagrado's number and from ergs to calories by dividing by 4.185×10^7, one can calculate the electrical work in calories per mole for charging up a cylinder of length *L*. Since *D* κ is practically constant between 10 and 25°C, W_{el} is almost independent of temperature in the range covered by our studies. Since W_{el} is known for a rod of length *L* and charge per protein subunit, *z*, it is possible to calculate ΔW_{el} for the reaction between two such rods to form one of double the length. ΔW_{el} is W_{el} for the longer rod minus twice W_{el} for the shorter ones. The results of calculations for various values of *z* and of the ratio, $L/(R + 1/2\kappa)$ are shown in Table 2.

Table 2. $+\Delta W_{el}$ for polymerization of TMV protein rods

L, Å	$\frac{L}{R + 1/2\kappa}$	$+\Delta W_{el}$, kcal/mole			
		$z = \pm 1$ (*pH*=5.2)	$z = \pm 2$ (*pH*=5.9)	$z = \pm 3$ (*pH*=6.6)	$z = \pm 4$ (*pH*=7.5)
49	1/2	3.8	15.2	34.2	61
98	1	8.6	34.4	77.5	137
196	2	13.7	54.8	123	229
392	4	17.0	68	153	272
784	8	19.2	77	173	307
∞	∞	21.1	84.5	190	338

The calculations in Table 2 indicate that the change in electrical work for polymerization of rods is positive, as one would expect it to be, that it increases with increasing negative charge and that it increases with increasing length of the polymerizing unit. Qualitatively, all of these results are of the right sort to explain the fact that the limiting length of rods capable of further polymerization decreases as *pH* and, therefore, net negative charge, increase. The question remains whether the values are of the correct order of magnitude to have a detectable effect on polymerization. More will be said about this in Chap. 5.

The Effect of Ionic Strength

The effect of ionic strength at various constant *pH* values on rod-forming polymerization of TMVP was measured by KHALIL and LAUFFER (16) and by SMITH and LAUFFER (2). In Fig. 7, data of KHALIL and LAUFFER show the variation in characteristic temperature as previously defined, T^*, with ionic strength, μ. The data are plotted for protein solutions in phosphate buffers in water at various *pH* values and in D_2O at various *pD* values. In all cases, the characteristic temperature is lower for higher values of ionic strength. $dT^*/d\mu$ was -56 in heavy water and -59 in water. ΔF^o_{290} values at particular values of μ were averaged for *pH* values of 6.0, 6.2, 6.4 and 6.6. In Fig. 8, $-\overline{\Delta F^o_{290}}$ is plotted against μ for data obtained in aqueous solutions. In the range between ionic strength 0.05 and 0.12, the value for $d\Delta F^o_{290}/d\mu$ is about -100 kcal/mole per mole per liter. A similar treatment of data obtained with heavy water gave a value of about -60.

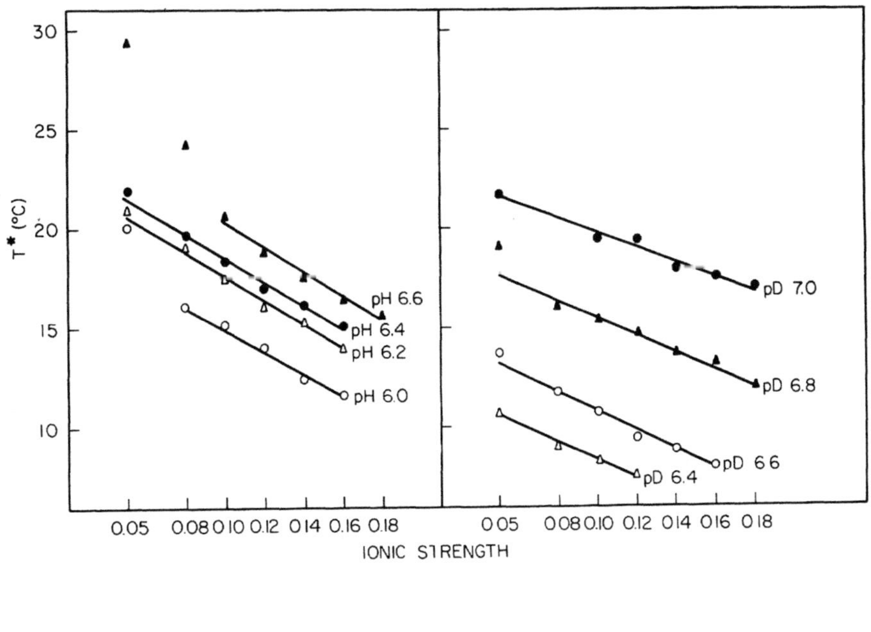

Fig. 7. Characteristic temperature (T^*) in degrees centigrade plotted against ionic strength. Data of KHALIL and LAUFFER (17). Reprinted from LAUFFER and STEVENS (18). Reprinted by permission of the copyright owner, Academic Press, Inc.

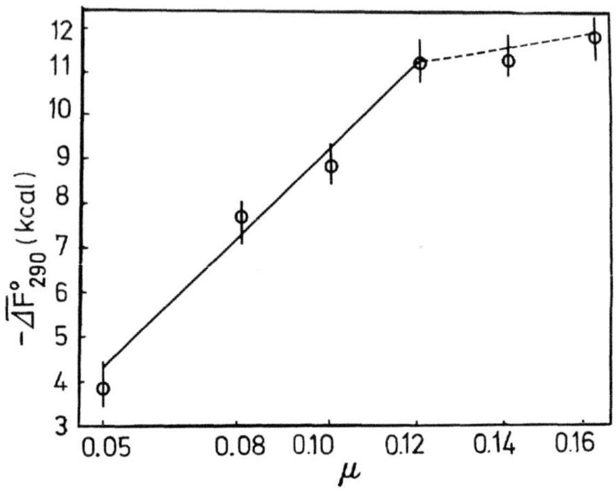

Fig. 8. Mean standard free energy for polymerization of TMVP plotted against ionic strength. Data of KHALIL and LAUFFER (16)

One way to interpret the effect of increasing ionic strength on polymerization of TMVP is in terms of the salting-out effect. If an organic solute is added to water, it decreases the dielectric constant, thereby increasing the attraction between oppositely charged ions present in the system. The result is a lowered solubility of the salt, which can be interpreted as an increase in the activity coefficient of the salt. The reciprocity principle of thermodynamics leads to the conclusion that if adding an organic solute to a solution increases the activity coefficient of the salt, adding salt must also increase the activity coefficient of the organic material. It is a fact that salts do increase the activity coefficient of organic materials dissolved in water. This is the salting-out effect. The salting-out of organic molecules is a reasonable analogue for the polymerization of TMVP, because this polymerization results in the transfer of an organic surface from an aqueous environment to an organic environment. If ions affect the activity coefficient of organic surfaces exposed to water but not those exposed to an organic environment in a "bond", the equilibrium constant, K, for polymerization must be replaced by $K_a f^2$, the forward reaction being bimolecular (15). K_a is defined as the equilibrium constant in terms of activity and f is the activity coefficient of the protein. The salting-out constant, K_s', is defined as the ratio of $\log_{10} f$ to ionic strength, μ. If, in the thermodynamic relationship between ln K and ΔF°, one substitutes $K_a f^2$ for K, one obtains Eq. (12).

$$-RT \ln K_a - 2 \times 2.3\, RT \log_{10} f = \Delta F^{\circ} \tag{12}$$

When $K_s' \mu$ is substituted for $\log_{10} f'$ and the equation is differentiated, one obtains $d\Delta F^o/d\mu = -4.6\ RT\ K_s' = 2660\ K_s'$ at 290°K. From the derivative of ΔF^o with respect to ionic strength of -100,000 cal/mole per mole per liter evaluated above, the value of K_s' can be calculated to be 37. Other ways of analyzing these data yield different values of the same order of magnitude. This salting-out constant is considerably higher than that found for most proteins. A value of 23 can be calculated for the results obtained with heavy water as the solvent.

The Effect of Various Chemicals

The effect of potassium thiocyanate, thiourea and acetamide on rod-forming polymerization of TMVP was studied by SHALABY and LAUFFER (17). Studies were carried out at several *pH* values and at several values of ionic strength. The results obtained with various concentrations of acetamide at four different *pH* values all at ionic strength of 0.1 are shown in Fig. 9. Here the characteristic temperature, T^*, is shown plotted against the molar concentration of acetamide. It is evident that at all *pH* values the characteristic temperature for polymerization is increased when acetamide is added. The straight lines shown in Fig. 9 can be described by intercept, a) equal to T^* when $m = 0$, and slope, b) equal to dT^*/dm. The data can then be described by recording a) and b), as is done in Table 3. Similar results were obtained with thiourea; in this case experiments were carried out at different ionic strength values as well as at different *pH* values. Results are shown in Table 3. Data comparable to those obtained with thiourea were also obtained with potassium thiocyanate. These are also shown in Table 3. In this case, the ionic strength values represent total values derived from buffer plus thiocyanate. In all three cases, the added chemicals increased the characteristic temperature for polymerization. That is, all three inhibited polymerization or favored depolymerization. Other chemicals were also studied by SHALABY and LAUFFER (17) and by SMITH and LAUFFER (2). EDTA, urea, dioxane and tetra-n-butylammonium bromide also favor depolymerization. Sucrose favors polymerization.

The effect of substituting thiocyanate ions for those in the buffer can be understood in terms of the salting-out effect. Thiocyanate is generally regarded as being less effective as a salting-out agent than many other negative

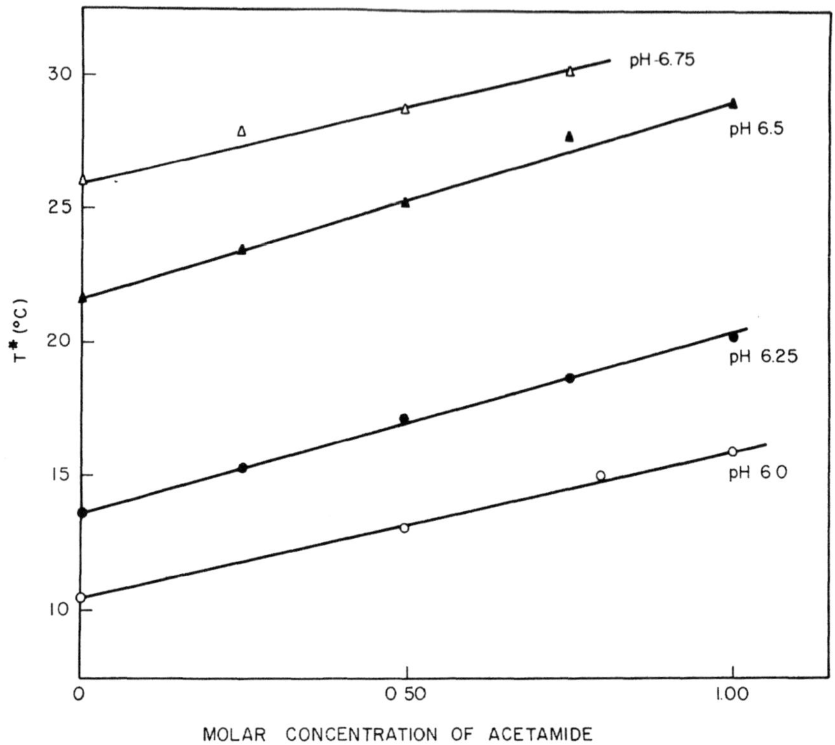

Fig. 9. Characteristic temperature for polymerization of TMVP (*T**) plotted against molar concentration of acetamide at $\mu = 0.1$. Data of SHALABY and LAUFFER (17). Reprinted from LAUFFER and STEVENS (18). Reprinted by permission of the copyright owner, Academic Press, Inc.

ions. All of those other chemicals which inhibit polymerization of TMVP are regarded as structure-breakers; sucrose, which favors polymerization, is regarded as a structure-maker. These terms have to do with the effect of the chemicals on the structure of water. The fact that reagents known to alter the structure of water influence the polymerization of TMVP is further evidence in favor of the view that release of water molecules is the principal source of the entropy increase required to drive the polymerization reaction. The fact that substitution of D_2O for H_2O enhances or facilitates polymerization fits this same general pattern, because D_2O has greater structure than H_2O.

Table 3. Values of a (T^* at $m = 0$) and b (dT^*/dm)

In acetamide

μ	0.10	
pH	*a*	*b*
6.0	10.4	5.5
6.25	13.6	6.8
6.5	21.6	7.4
6.75	26.0	5.5

In thiourea

μ	0.125		0.10		0.075	
pH	*a*	*b*	*a*	*b*	*a*	*b*
6.0	8.4	15.7	9.0	16.2	10.4	16.7
6.25	13.2	16.6	13.8	15.3	15.6	18.3
6.5	18.4	23.7	----	----	21.9	28.5
6.75	24.8	20.7	25.5	27.5	----	----

In KSCN

μ	0.15		0.125		0.10		0.075	
pH	*a*	*b*	*a*	*b*	*a*	*b*	*a*	*b*
6.0	----	----	----	----	----	----	----	----
6.25	11.8	40.8	12.9	28.0	13.6	28.8	16.0	20.0
6.5	18.5	57.0	18.5	44.0	21.3	60.0	22.8	60.0
6.75	26.2	35.0	25.5	68.0	27.3	46.7	27.3	56.0

Calorimetric Studies

The rod-forming polymerization data obtained by the method of light scatter all involved experiments in which *pH* was maintained constant by buffer and the extent of polymerization of TMVP was changed by changing temperature. The ΔH^{o} values calculated from these data are van't Hoff enthalpies. When measurements are made in the calorimeter, polymerization is brought about by other means, such as changing *pH*, ionic strength or protein concentration. The most extensive calorimetric studies of the heats absorbed in rod-forming polymerization of TMVP are those of SRINIVASAN and LAUFFER (23). Their data are summarized in Fig. 10. The filled circles represent TMVP in 0.01 μ buffer, and the triangles represent TMV in similar solutions. For the case of TMVP, each experimental point represents the heat absorbed for a hypothetical experiment in which the protein at a concentration of 2.5 mg/ml was transferred from a 0.01 μ phosphate buffer at *pH*=7.5 to a buffer at the same ionic strength at the *pH* value indicated on the graph, without change of protein concentration. In all of these experiments, thermal effects attributable to changes in buffer composition were subtracted automatically, because a twin vessel calorimeter was used, one vessel of which contained protein solution in buffer and the other buffer alone. The recorded result was always the difference between the two. Actually, in the experiments there were changes in protein concentration, but corrections were made to constant concentration by using separate data on the heat changes realized by diluting the protein at constant *pH* and constant temperature. The data for TMV are comparable, with two exceptions. Firstly, they represent the heats absorbed per gram of protein in the virus, not per gram of virus. Second, for reasons which will be explained below the graph was displaced upward so that it coincides with the graph for TMVP in the region between *pH*=3.5 and 5.0. Similar results were obtained for the protein in 0.1 μ buffers, except that the steep rise in heat absorption found at *pH*=6.3 on the graph occurs at *pH*=6.5. There is no change of heat content when TMV is transferred from 0.1 μ phosphate buffer to 0.01 μ buffer. Thus, the results shown in Fig. 10, with the exception noted, apply equally to reaction in 0.1 μ phosphate buffer. All experiments were carried out at room temperature, approximately 23°C.

Between *pH*=7.5 and 6.3, very little heat is absorbed because there is very little change in the degree of polymerization of the protein. At about *pH*=6.3 there is a very sharp increase in heat absorption. This is brought about by

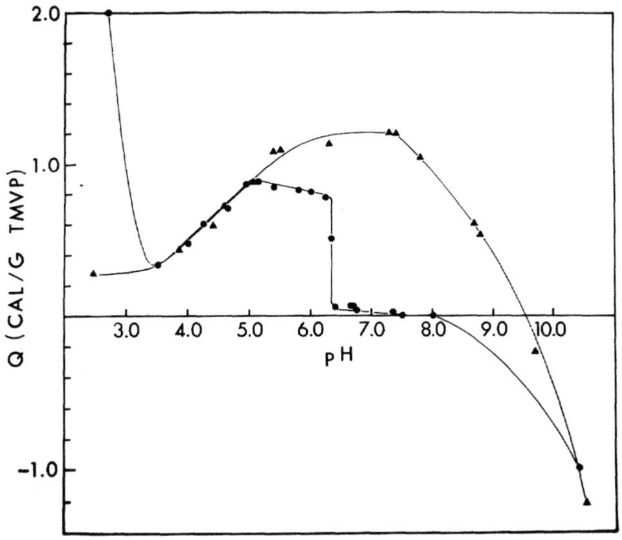

Fig. 10. Plot of Q (cal/g of protein) as a function of *pH*. ▲, TMV in 0.1 μ phosphate buffer; ●, TMVP in 0.01 μ phosphate buffer. From SRINIVASAN and LAUFFER (23). Reprinted by permission of the copyright owner, The American Chemical Society

the polymerization of the protein at this *pH* value. Between *pH*=6.3 and 5.0 there is a very gradual increase in heat absorbed. This correlates well with the fact observed by JAENICKE and LAUFFER (24) that there is a small increase in partial specific volume when the *pH* is reduced from 6.3 to 5.0. The explanation for both phenomena is that the structure of the polymerized protein changes from a rather loose to a more compact one over this *pH* range.

At *pH*=5.0, crystallographic studies showed that the structure of the polymerized TMVP is indistinguishable from the structure of the protein component of TMV. Both TMV and TMVP exhibit a decrease in Q when the *pH* is reduced from 5.0 to 3.5. Hydrogen ion titration curves show that the number of hydrogen ions bound in this *pH* region is about the same for protein and virus. Therefore, this region of the curve is attributed to hydrogen ion binding for both TMVP and TMV. This is the reason for drawing the curve obtained for the TMV data to coincide in this *pH* region with that for the TMVP data. The difference between the value of Q obtained for TMV at *pH*=3.5 and that obtained at *pH*=7.5 agrees very well with the values calculated when it is assumed that all thermal effects with TMV in this *pH* region are the result of the change with *pH* in the number of hydrogen ions bound per protein subunit. Therefore, the TMV curve represents the *Q* value one would obtain with TMVP at a particular *pH* value if the structure of the polymerized TMVP at that *pH* value were the same as the protein arrangement in TMV or as the protein arrangement in both TMV and TMVP at *pH* values between 5.0 and 3.5. One can therefore estimate the heat absorbed in a hypothetical polymerization reaction at *pH*=7.5, in which unpolymerized protein is polymerized at constant *pH* into a compact structure like that found in TMV. This turns out to be 1.2 cal/g. Practically the same result would be obtained at any *pH* values between 6.5 and 7.5. This result can be converted to calories per mole of polymerizing substance if the molecular weight of the starting material is known. Osmotic pressure results (23) indicate that a value of 107,000 is a good estimate for the number average molecular weight at a concentration of 2.5 mg/ml. For the purpose of this calculation, Eq. (12) of Chap. 3 can be used. Since the value of the molecular weight in the highly polymerized state is very much larger than that in the unpolymerized state, the equation can be similified to equate ΔH to M_1Q. Therefore, the enthalpy of this hypothetical polymerization reaction is 107,000 times 1.2, or 129,000 cal/mole. If the view we have adopted is correct, namely that cyclical trimers and integral polymers of cyclical trimers first form double spirals (Reaction I) and double spirals then polymerize (Reaction II) to form helical rods, the enthalpies thus calculated would represent average enthalpies for the two steps.

It is possible to use the same data to obtain an estimate of the enthalpy of rearrangement of double discs and polymerization to a structure comparable to that found in the virus at *pH*=5, Reaction IV and II. From the data of STAUFFER *et al.* (25), one can interpret that if TMVP is transferred from 0.01 μ phosphate buffer at *pH*=7.5 at a concentration of 2.5 mg/ml to 0.1 μ phosphate buffer at *pH*=7.5 at a concentration of 50 mg/ml, 0.325 cal will be absorbed per gram of protein. All available evidence indicates that the protein exists in the form of double discs under the latter condition. Since no heat is evolved or absorbed when TMV is transferred from 0.1 μ to 0.01 μ buffers, it is assumed that the same is true for polymerized protein. Therefore the heat absorbed in polymerizing one gram of double discs in 0.1 μ phosphate buffer at *pH*=7.5 and rearranging into a structure comparable to that of protein in virus should be 1.2 - 0.325, or 0.875 cal/g. When this figure is multiplied by the molecular weight of double discs, 596,000, the enthalpy for polymerization of double discs is calculated to be 520 kcal/mole. This hypothetical polymerization involves no change in charge. It should be comparable to real polymerizations which take place at pH values where there is no change in the charge, for example at pH values below 6. KHALIL and LAUFFER (16) and SHALABY and LAUFFER (17) reported van't Hoff enthalpies for polymerization in 0.1 μ phosphate buffer at pH=6 ranging between about 400 and 700 kcal/mole.

The agreement between the calorimetric enthalpy thus calculated and the van't Hoff enthalpy is better than the internal consistency of the various estimates of the van't Hoff enthalpy. It is actually better than one has any right to expect. Calorimetric and van't Hoff enthalpies frequently differ. The usual way of explaining deviations of this sort is in terms of the Kirchoff equation, presented as Eq. (6) of Chap. 1. This equation points out that ΔH might not be a constant independent of temperature, but might vary with temperature because the reactants and the products of the reaction have different heat capacities at constant pressure. This equation can be written in the form

$$(d\Delta H^{\circ})/d(1/T))_{p} = -\Delta C_{p}\, T^{2} \quad (13)$$

Since ΔH° is now regarded as a variable dependent on $1/T$, the derivative of Eq. (3) of Chap. 1 now becomes;

$$-Rd(\ln K)/d(1/T) = \Delta H^{\circ} - T\Delta C_{p} \quad (14)$$

For the kind of reaction being considered here, ΔC_p means the difference between the molar heat capacity of polymerized protein multiplied by the ratio of the molecular weight of unpolymerized protein to that of polymerized protein and the molar heat capacity of the unpolymerized material. In this equation as finally written, ΔH° should correspond to the true calorimetric enthalpy. No information is currently available concerning the magnitude of ΔC_p for the polymerization of TMVP.

Summary

At *pH* values below 7, except at low temperatures in the neighborhood of *pH*=6.5, TMVP polymerizes to form rods in which the protein chemical subunits are arranged in the helical manner in which they are found in TMV. For historical reasons, this type of polymerization was described as "high temperature" polymerization, because at *pH*=6.5 and 0.1 ionic strength phosphate buffer, this reaction proceeds at temperatures above about $14^{\circ}C$, in contrast to the fact that at temperatures between $4^{\circ}C$ and $14^{\circ}C$, the type of polymerization described in Chap. 3 occurs. In the present chapter, what was formerly called "high temperature" polymerization is referred to as rod-forming polymerization. Most of what we know about the thermodynamics of this kind of polymerization has been derived from light scatter studies. The data thus obtained can be interpreted from several different points of view. The interpretation originally proposed by the author's laboratory is that small units, probably cyclical trimers of the basic chemical unit, polymerize in a manner comparable to condensation polymerization. It now seems probable that light scatter used to follow rod-forming polymerization is sensitive to the polymerization of double discs or double spirals of comparable molecular weight. This assumption also permits the data to be interpreted in terms of the mathematics of condensation polymerization. From the data thus analyzed, van't Hoff enthalpies can be calculated. Another way of looking at the data is in terms of cooperativity. Many aspects of rod-forming polymerization bear at least superficial resemblance to aspects of phase transitions, typical cooperative processes. However, a particular stage of a process cannot at the same time be typically cooperative and follow the laws of condensation polymerization, because the basic assumption underlying cooperative processes in that the reactivity of a unit already combined is different from that of the unit in the uncombined state. A polymerizing substance can follow the laws of condensation polymerization only when the reactivity of a unit is independent of whether it is combinded or free.

In the *pH* range between 7 and 6, the effect of lowering *pH* is to increase the van't Hoff enthalpy, the entropy and the negative free energy of polymerization and to increase the maximum extent of polymerization. The maximum extent of polymerization can be understood in terms of diminished electrostatic repulsion in the rods ultimately formed when the products obtained in the early stages polymerize further.

When ionic strength is increased, rod-forming polymerization is enhanced. This can be interpreted quantatively in terms of salting-out theory. The effect of several chemicals on rod-forming polymerization of TMVP has been studied. Thiocyanate ion increases the temperature at which polymerization takes place; that is, it favors depolymerization. This can also be understood in terms of the salting-out effect, because thiocyanate is a less effective salting-out agent than many other negative ions. EDTA, urea, thiourea, acetamide, dioxane and tetra-n-butyl ammonium bromide also favor depolymerization. Sucrose and heavy water favor polymerization. Those chemicals which inhibit polymerization are generally regarded as breakers of the structure of water; sucrose is regarded as a structure-maker and D_2O has greater structure than H_2O.

Finally, direct measurements of the enthalpy of polymerization of TMVP show that the calorimetric enthalpy is about 520 kcal/mole of double spiral, a value within the range for van't Hoff enthalpies found by light-scatter studies.

Chapter 5

Polymerization-Depolymerization of the Proteins from Mutants of TMV and Other Viruses

Introduction

Studies on the polymerization-depolymerization of the proteins isolated from cucumber virus 4, the Dahlemense strain of TMV and the PM2 mutant of TMV have been carried out in the author's laboratory. The proteins of these three viruses are similar to TMVP, but differences in amino acid composition and sequence exist. Polymerization of the protein from Cucumber Green Mottle Mosaic Virus, related to TMV, and Tobacco Rattle Virus have been studied in other laboratories.

Cucumber Virus 4

Cucumber virus 4 (CV4) is morphologically similar to and serologically related to TMV (1, 2). These facts suggest that CV4 is a strain of TMV. However, KNIGHT (3) recommended that it not be thus classified because of the extent of chemical and biological difference. For our purposes, this is not an important issue.

The normal host of CV4 is the cucumber; the virus can be isolated from infected cucumber plants in a manner similar to that used for isolation of TMV from infected tobacco plants. FRIST *et al.*, (4) prepared the nucleic acid-free protein from CV4 by the acetic acid method used in the preparation of TMVP. This protein (CV4P), like TMVP, exhibits reversible polymerization into virus-like rods at *pH* values near 6. In 0.1 μ tris-saline buffer at *pH*=8.0, CV4P at concentrations of 6 mg/ml and higher has a sedimentation coefficient at 4°C of about 3 Svedbergs. This is significantly lower than the corresponding value for TMVP; it suggests that the basic mode of polymerization of CV4P involves the formation of dimers of the chemical unit as the first step. Like TMVP, CV4P dissociates into more slowly sedimenting components upon dilution to a concentration of 1 mg/ml or less.

The amino acid composition of CV4P shows many similarities to but some significant differences from that of TMVP, (5,6). It has only 151 amino acid residues, in contrast to 158 for TMVP. The subunit molecular weight is 16,102. In common with TMVP, CV4P has no histidine and no methionine. It has considerably more serine and considerably more alanine and one more lysine. The most conspicious difference is the absence of cysteine; TMVP has one.

FRIST *et al.*, (4) reacted CV4P with succinic anhydride. The reaction was carried out at 5°C with constant stirring, and the *pH* of the solution was maintained between 7 and 8 by periodic addition of sodium hydroxide. After succinylation, the preparations were dialyzed against buffer. Sedimentation, diffusion and osmotic pressure studies were carried out with the results shown in Figs. 1, 2 and 3. From the extrapolated values of the sedimentation and diffusion data, the molecular weight of the succinylated CV4P was calculated to

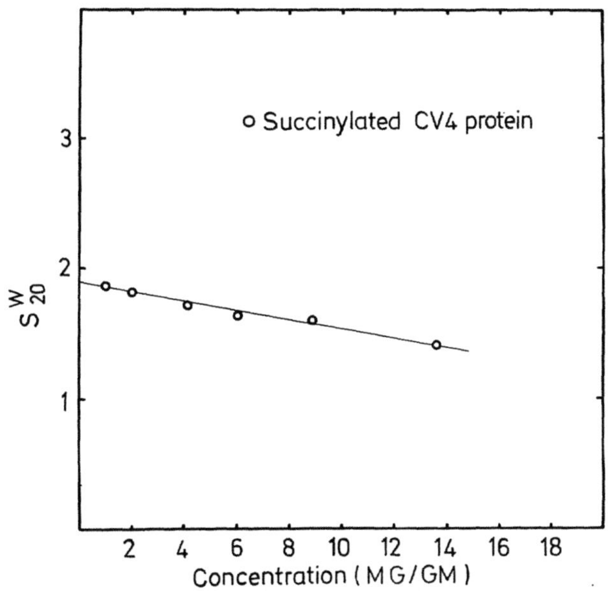

Fig. 1. Sedimentation of succinylated CV4 protein. Drawn from data of FRIST *et al.*, (4)

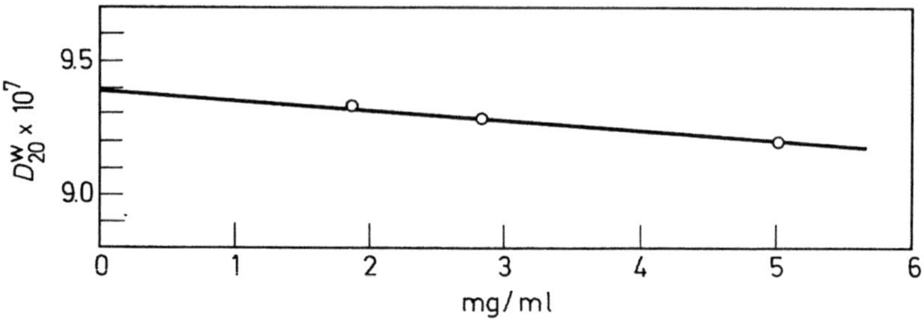

Fig. 2. Diffusion of succinylated CV4 protein. D^{w}_{20} is expressed in $cm^2\ sec^{-1}$. From FRIST *et al.* (4). Reprinted by permission of the copyright owner, Academic Press, Inc.

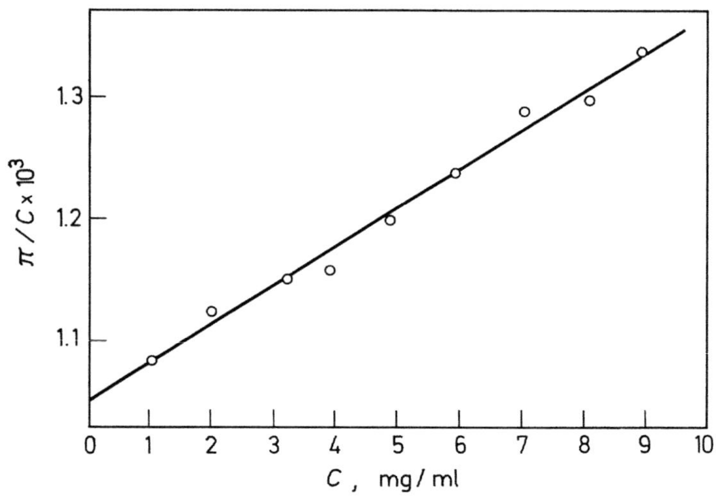

Fig. 3. Osmotic pressure data for succinylated CV4 protein. π is expressed in atmospheres. From FRIST *et al.* (4). Reprinted by permission of the copyright owner, Academic Press, Inc.

be 19,000. The osmotic pressure data, extrapolated to zero concentration, yielded a molecular weight of 21,500. Since chemical analysis of CV4P indicate a subunit with molecular weight of about 16,102 (5,6), these molecular weight determinations show that succinylation causes the protein to dissociate into its monomeric form.

Succinic anhydride reacts with lysine residues, thereby substituting a negatively ionizing bulky group for a small positively ionizing amino group. This results in a large net increase in the negative charge of the protein, which should be reflected in an increase in electrophoretic mobility at *pH*=7. Furthermore, since CV4P seems to have three lysines (5) while TMVP has only two per protein subunit, the effect of succinylation on the electrophoretic mobility of CV4P should be greater than that on TMVP. Electrophoretic mobility measurements at $4^{o}C$ in 0.1 μ Michaelis buffer at *pH*=7 showed an increase in negative electrophoretic mobility upon succinylation of CV4P from 2.8 to 11.4×10^{-5} cm/sec/per volt/cm. Comparable changes upon succinylation of TMVP were from 3.9 to 10.8 (4).

The hydrogen ion titration curve at $4^{o}C$ in the region where polymerization takes place, shown in Fig. 4, is very similar in shape to that of TMVP at $4^{o}C$, as shown in Fig. 8 of Chap. 4. Quantitatively, there is a difference. When the *pH* is lowered from 7.5 to 5.5, the range over which protein changes from unpolymerized at all temperatures to polymerized at all temperatures, TMVP binds three and CV4P binds four hydrogen ions per chemical monomer. This suggests that there might be a greater binding of hydrogen ions by CV4P when polymerization is carried out at constant *pH* by raising the temperature. FRIST (7) compared the electrophoretic mobility at various *pH* values in 0.1 μ Michaelis buffer at 4° of CV4 and CV4P with TMV and TMVP, as shown in Fig. 5. The results on TMV and TMVP were similar to those reported by KRAMER and WITTMANN (8). At *pH* values less than 6, polymerized TMVP had a mobility identical to that of TMV, but when TMVP was depolymerized at *pH* values greater than 6, its mobility was significantly lower than that of TMV. Thus, the mobility versus *pH* curve for TMVP shows a discontinuity when protein passes from the polymerized to the unpolymerized form. When TMVP disaggregates at *pH* values above 6, charge per unit mass increases because of the dissociation of hydrogen ions. But the surface area per unit mass increases greatly. Electrophoretic mobility depends on the ratio of charge to surface area. In the case of TMVP this ratio decreases, resulting in a big drop in mobility upon depolymerization.

In contrast, the mobility of CV4P followed that of CV4 over the entire range of *pH* from 5.5 to 8.5. No significant change in mobility was observed when CV4P depolymerized near *pH*=6. The following argument shows that this result is consistent with hydrogen ion titration data. If the number of hydrogen ions dissociated per unit of mass is greater for CV4P than for TMVP, then the

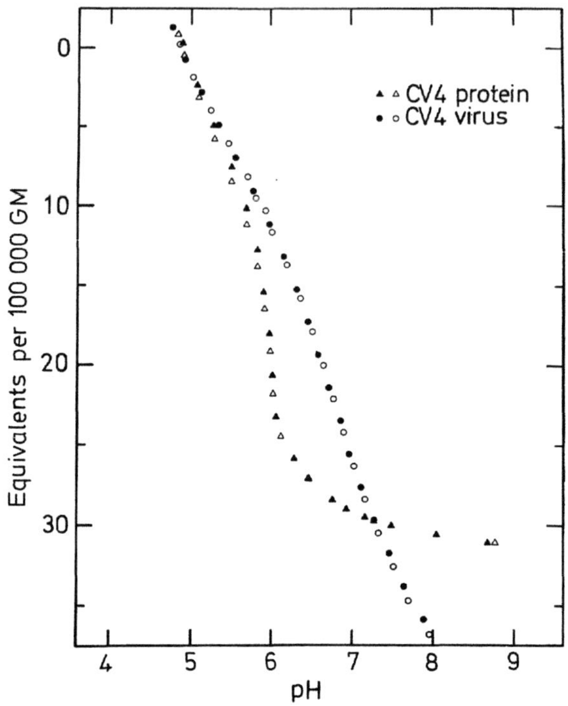

Fig. 4. Acid-base titration curve of unbuffered CV4 protein and unbuffered intact virus in 0.1 μ NaCl at $4^{o}C$. Open symbols represent addition of base and solid symbols subsequent addition of acid. From FRIST (7)

increase in charge per unit mass upon depolymerization will be greater for CV4P than for TMVP. Since the increase in surface per unit of mass should be approximately the same for the two, the change in the ratio of charge to surface area should be less in the case of CV4P than in the case of TMVP. When this is combined with the fact that even polymerized CV4P at *pH*=6 has a very low mobility compared to that of TMVP, about one fourth, the change turns out to be undetectable.

Polymerization results on CV4P, reported briefly by FRIST *et al.* (4) and more fully by FRIST (7), show that in 0.1 μ phosphate buffers, at *pH* values between 6.75 and 8.0, the protein is unpolymerized at all temperatures between 4 and $25^{o}C$. At *pH* values between 5.5 and 5.75, the protein is polymerized at all temperatures studied. Between *pH*=5.75 and 6.75, these solutions are polymerized at higher temperatures and unpolymerized at lower temperatures. The

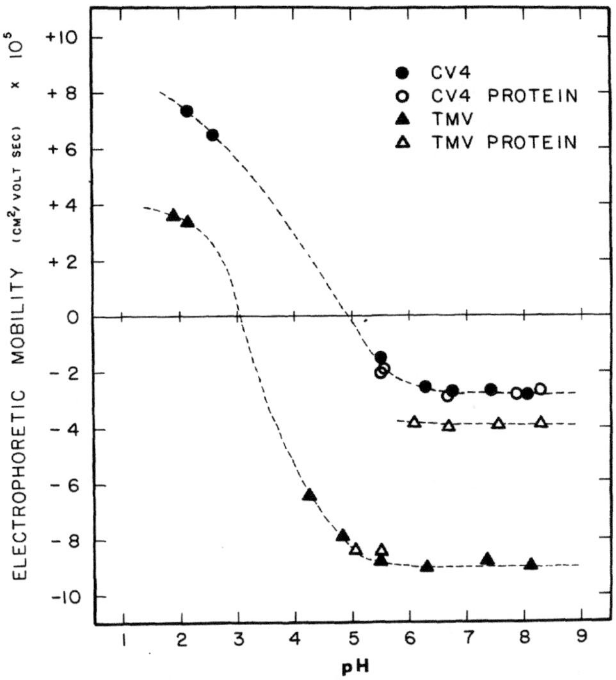

Fig. 5. Electrophoretic mobility in 0.1 μ Michaelis buffer at 4°C of CV4 and TMV and their proteins. From FRIST (7)

greatest change occurred at *pH*=6.25. Just as in the case of TMVP, the process at *pH*=6.25 is completely reversible. It was followed both by light scatter and by flow birefringence; the observations were qualitative rather than quantitative. Nevertheless, the polymerization pattern is very similar to that obtained with TMVP.

Sedimentation results showed that, at *pH*=8 in 0.1 μ buffer, CV4P had a sedimentation coefficient, s_{20}^{w}, of about 3 S. At *pH*=5.5 the same protein had a sedimentation coefficient, s_{20}^{w}, of about 140 S (7). In some cases, particularly when the unpolymerized protein was transferred to an acetate buffer rapidly instead of by dialysis, a more slowly sedimenting component with a sedimentation coefficient of 25-30 S was detected. Fig. 6 is an electron micrograph of such a preparation which showed two peaks with coefficients of 138 S and 26 S in the ultracentrifuge. It seems to be composed of a mixture of double discs or very short stacks of double discs and long particles with the appearance of helical rods. Except for length, these particles were indistinguishable in appearance from CV4 particles.

Fig. 6. Electron micrograph of CV4 protein. Prior to preparation of the mount for electron microscopy, the protein had been brought to *pH*=5.5 rapidly by adding acid to a solution in 0.1 μ buffer at *pH*=8. From FRIST (7)

Overall, the indications are that the polymerization behavior of CV4P is very much like that of TMVP. At low temperatures in neutral solution, CV4P exists in the form of slightly polymerized material, possibly dimers of the ultimate protein monomer rather than the trimers found in the case of TMVP. When the *pH* is lowered to 5.5, even at low temperatures, the protein polymerizes into long helical rods with double discs or something closely resembling them as intermediates. At *pH* values in the neighborhood of 6.25 to 6.5 in 0.1 μ buffer, the protein polymerizes reversibly when temperature is raised. The change, as in the case of TMVP, is accompanied by binding of hydrogen ion. The indications are that more hydrogen ions are bound by CV4P than by TMVP.

Treatment of protein with succinic anhydride results in the replacement of an NH_2 group with an $NHCO(CH_2)_2COOH$, or it converts a positively ionizing center into a negatively ionizing center. This is the explanation for the very big increase in negative electrophoretic mobility upon treatment with succinic anhydride of both TMVP and CV4P. This increase in negative charge apparently causes dimers or trimers to dissociate at least partially into monomers. It also prevents polymerization of both CV4P and TMVP into rods when *pH* is lowered.

The Dahlemense Strain

The Dahlemense (DTMV) strain of TMV was isolated from diseased tomato plants by MELCHERS (9). It is regarded as a strain of TMV because of similarities in size, shape, composition and structure and because of serological relationships.

The coat-protein (DTMVP) from the Dahlemense strain can be isolated from the virus by methods similar to those used for TMVP. The chemical repeat unit of DTMVP is similar to that of TMVP in that it contains 158 amino acids. By comparing the amino acid sequence obtained by WITTMANN-LEIBOLD and WITTMANN (10) for DTMVP with that for TMVP shown in Fig. 1 of Chap. 3, one finds 28 differences, including one fewer aspartic acid residue, two fewer arginine residues and one methionine residue compared to none for TMVP. Thus, the net negative charge near *pH*=7 on DTMVP should be one unit higher than for TMVP. As shown in Fig. 7, electrophoretic mobility data obtained by KRAMER and WITTMANN (8) on polymerized TMVP and DTMVP in 0.075 μ Michaelis buffer are in accord with this difference in number of potential positive ions. The isoelectric points of DTMVP and TMVP are at *pH*=2.7 and 3.2, respectively, and the maximum negative electrophoretic mobilities near *pH*=6, the highest *pH* value at which the polymerized form exists at 4°C, are 0.82×10^{-4} cm/sec per volt/cm for TMVP and 0.92×10^{-4} for DTMVP. Furthermore, the mobilities of the polymerized protein coincide with the mobilities with the respective viruses. TMVP and DTMVP can co-polymerize into rods with electrophoretic mobility intermediate to those of polymerized TMVP and polymerized DTMVP, (11).

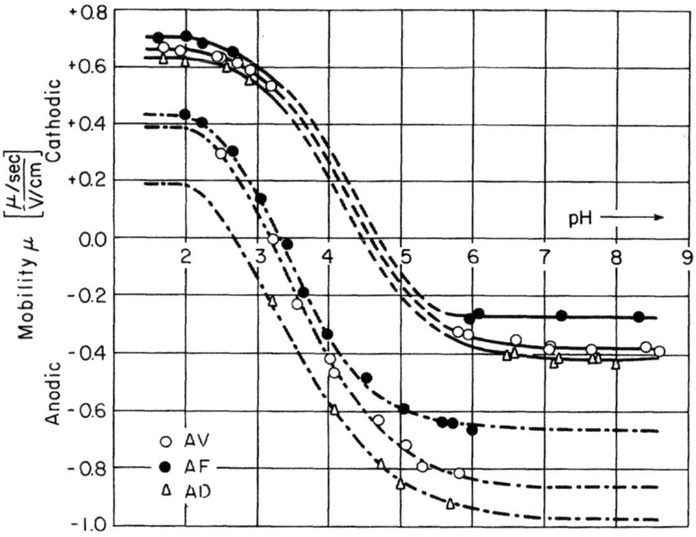

Fig. 7. Electrophoretic mobility of proteins from three strains of TMV (vulgare, AV; flavum, AF: dahlemense, AD) as a function of *pH* in 0.075 μ buffer at 4°C. From KRAMER and WITTMANN (8). Reprinted by permission of the copyright owner, Zeitschrift für Naturforschung

The X-ray diffraction pattern of DTMV is similar to that of TMV in showing a helix with a 23 Å pitch containing 49.02 protein subunits in 3 complete turns (12). However, one important difference, illustrated in Fig. 8, was observed. In TMV, shown on the right (Fig. 8b), the outer ends of the subunits are equivalently related, that is, all are at the same distance from nearest neighbors above and below. However in DTMV, shown on the left (Fig. 8a), the outer ends of the subunits in two turns of the helix are pinched together, necessitating a wider separation between the second and third turns. CASPAR and HOLMES (12) explained that the inner ends (closest to the cylinder axis) of the DTMV protein subunits form a perfectly regular helix with a 23 Å pitch. Along one line parallel to the axis, the outer edges of the protein subunits are spaced at 23 Å, the pitch of the helix. However, starting at this point, successive subunits have their outer edges bent downward more and more as one proceeds around the helix until maximum bending is achieved at 180°. Between 180° and 360°, the downward bending of the outer edges of successive protein subunits becomes less and less until it is zero at 360°. In the second turn, successive protein subunits have their outer edges displaced upward, beginning with zero displacement at 0°, exhibiting a maximum upward displacement at 180° and again zero at 360°. Thus, along the reference line chosen, there is no pinching together of alternate turns of the helix and at 180° to it, there is maximum pinching.

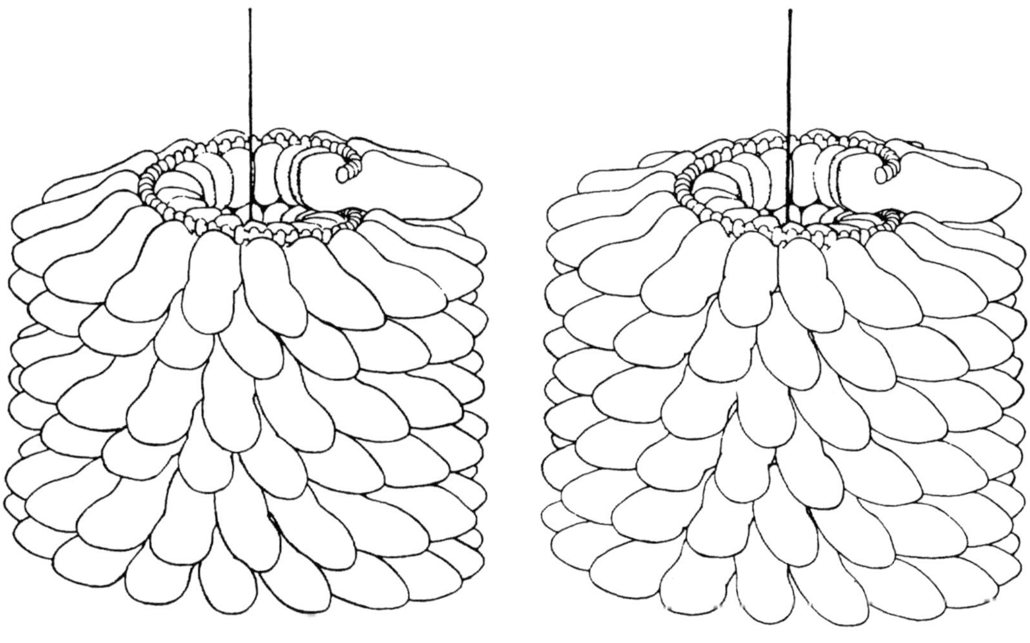

(a) (b)

Fig. 8a and b. Diagramatic comparison of the structure of the Dahlemense strain (a) and the common strain (b) of TMV. Helices are left-handed, contrary to modern view (Fig. 3 of Chap. 2). From CASPAR and HOLMES (12). Reprinted by permission of the copyright owner, the Journal of Molecular Biology

The polymerization of DTMVP has been studied by SRINIVASAN and LAUFFER (13). As illustrated in Fig. 9, when DTMVP is dissolved in 0.1 μ phosphate buffer at *pH*=6.25, its variation of degree of polymerization with temperature is practically indistinguishable from that of TMVP. At *pH*=6.0, DTMVP polymerizes at lower temperatures, and to a greater extent at a given temperature, than does TMVP. At *pH*=6.5, DTMVP polymerizes gradually and only to a small extent as temperature increases, while TMVP polymerizes abruptly with temperature increase. At *pH* values above 6.5, both DTMVP and TMVP polymerize gradually and to a small extent, but the extent is somewhat greater for DTMVP. Fig. 10 shows that in 0.05 ionic strength phosphate buffers, DTMVP does not polymerize at all at *pH*=6.25 and polymerizes to a much lesser extent than TMVP at *pH*=6.0. It is thus apparent that DTMVP is much more sensitive to changes in ionic strength than TMVP.

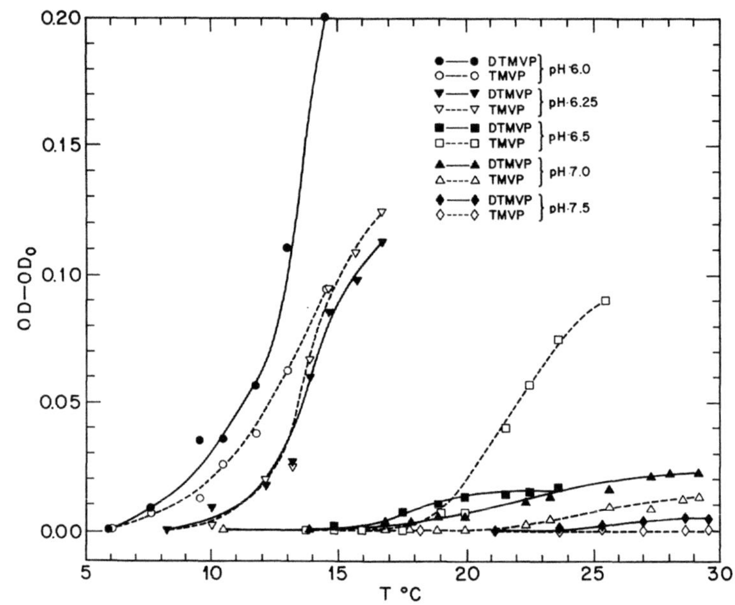

Fig. 9. Optical density (320 mμ) vs. temperature. Effect of *pH* on polymerization of DTMVP and TMVP in 0.1 μ phosphate buffer. From SRINIVASAN and LAUFFER (13). Reprinted by permission of the copyright owner, Academic Press, Inc.

SRINIVASAN and LAUFFER (13) have also obtained hydrogen ion titration curves on DTMVP at 4° and at 23°C, shown in Fig. 11. In the region just above *pH*=6, where the protein is unpolymerized at 4°C and highly polymerized at room temperature, the difference between the two titration curves is greater than the corresponding difference observed for TMVP. This shows that the number of hydrogen ions bound per chemical subunit of protein upon polymerization is greater for DTMVP than for TMVP. Prior information was available on the acid-

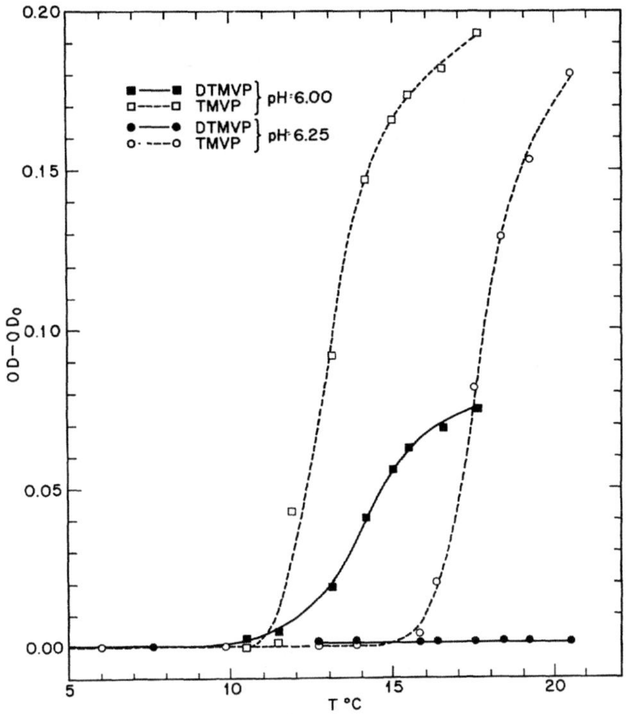

Fig. 10. Optical density (320 mμ) vs. temperature. Effect of *pH* on polymerization of TMVP and DTMVP in 0.05 μ phosphate buffer. From SRINIVASAN and LAUFFER (13). Reprinted by permission of the copyright owner, Academic Press, Inc.

base titration data on a number of strains of TMV, including the normal type and the Dahlemense strain. For all of them, 7.5 ± 0.5 hydrogen ions are bound between *pH*=6.5 and 1.8. While DTMV and TMV bind the same number of hydrogen ions over this range, the shapes of the two titration curves are considerably different. Acid-base titration results obtained by SCHEELE and LAUFFER (15) on TMV show 7 hydrogen ions bound between *pH*=6.5 and 2.5. Since in the protein repeat unit of TMV, there are 16 carboxyl groups, eight aspartic acid, seven glutamic acid and one terminal, and since the total number is 15 for DTMV, eight carboxyl groups in DTMV and nine in TMV titrate at *pH* values beyond the normal range. Allowing for the possibility that two carboxyl groups titrate at *pH* values far above the normal range (16, 17), this leaves six or seven which titrate at *pH* values below 1.8. OEHLEN (14) concluded that these must be in an undissociated state in all strains of TMV. The results of SCHEELE and LAUFFER obtained on TMVP, shown in Fig. 6 of Chap. 4, throw further light upon this. These groups do titrate between about pH=1.8 and pH=1. Actually, 17 hydrogen ions are bound between pH=7 and pH=1, one more than the total number of carboxyl groups present. If all the data are accurate, in addition to all sixteen carboxyl groups, one positive group which normally binds hydrogen ions at pH values above 7 must be titrated in the acid range.

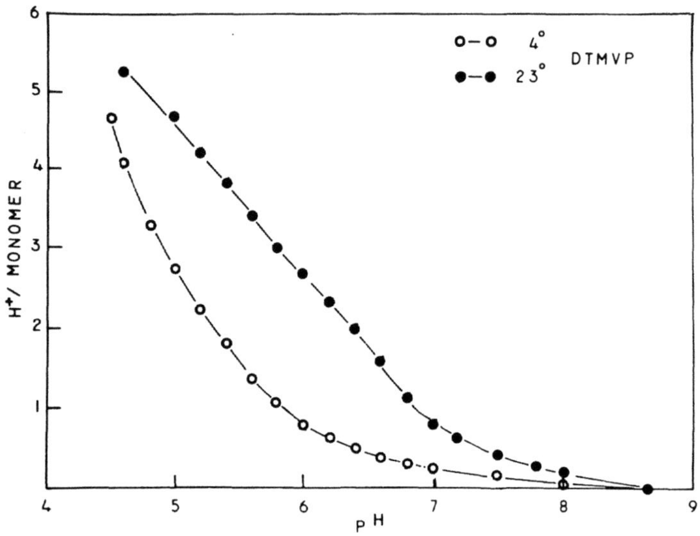

Fig. 11. Acid-base titration curves of DTMVP at 4°C and 20°C. From SRINIVASAN and LAUFFER (13). Reprinted by permission of the copyright owner, Academic Press, Inc.

Fig. 12 is an electron micrograph of DTMVP in 0.1 ionic strength phosphate buffer, pH=6.5 at a concentration of 0.1 mg/ml. The micrograph (a) on the left was obtained on a grid prepared from a solution at 4°C, and the micrograph (b) on the right was obtained on a grid prepared from a solution at 23°C. It is thus apparent that particles resembling double discs predominate at low temperature and that the polymerization process at pH=6.5 seems to be one involving the polymerization of small discs into long rods. Ultracentrifuge data on DTMVP in 0.1 μ phosphate buffer at temperatures ranging between 6°C and 12°C showed that all of the material sedimented at a rate of 34 S at pH=6.25 and 40% and 30% respectively, sedimented with a sedimentation coefficient of 23 S at pH=6.4 and 6.5. The balance sedimented approximately as a 4 S component. The 23 S component probably corresponds to double discs, and the 34 S component to slightly polymerized double discs. Osmotic pressure results

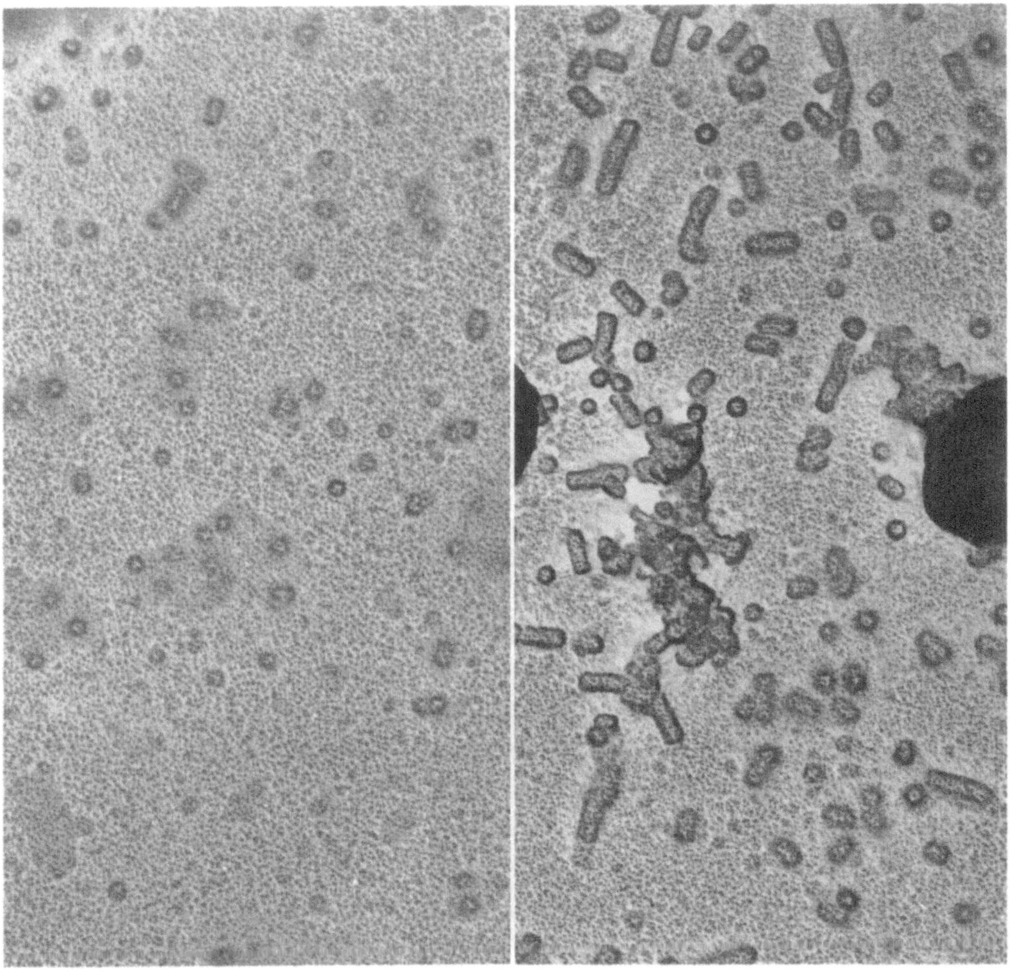

(a) (b)

Fig. 12a and b. An electron micrograph of DTMVP in 0.1 μ phosphate buffer, *pH*=6.5, concentration 0.1 mg/ml, (ca. 10^5x): (a) at 4°C; (b) at 23°C. From SRINIVASAN and LAUFFER (13). Reprinted by permission of the copyright owner, Academic Press, Inc.

obtained in 0.1 μ phosphate buffer at *pH*=6.4 and 7.5°C indicate a number average molecular weight of 554,000, again indicating that the state of the material in solution at low temperature approximates that of the double disc. The weight of the evidence indicates that the typical optical density vs. temperature curve obtained at *pH*=6.25 in 0.1 μ phosphate buffer, as illustrated in Fig. 9, represents the polymerization of particles with the same or somewhat higher molecular weight as double discs into very long rods. This is in contrast with the finding with TMVP reported in Chap. 3. In the case of TMVP in 0.1 μ phosphate buffer at *pH*=6.5, the protein is predominately in the form of smaller particles, presumably trimers, at temperatures in the neighborhood of $6^{O}C$. However, at *pH* values of 6.0 and 6.25, TMVP exhibits both 20 S and 4 S material in the ultracentrifuge at $4^{O}C$, (18).

The PM2 Mutant

The PM2 mutant of TMV was produced in the laboratory by SIEGAL *et al.*, (19) by treating wild-type virus with nitrous acid until the survival was only 0.015%. This virus infects tobacco plants, producing yellow spots which appear about a week after inoculation on mature leaves. The infection spreads throughout the leaf and sometimes to other tissues of the plant and to leaves which were not inoculated. A protein is synthesized as a result of infection, but the PM2 mutant is defective in that its protein does not coat its RNA to form complete virus particles. The infectious principle is thus naked RNA, which is replicated during the infection process and which induces the formation of a defective protein.

PM2 protein has been isolated and purified by a method involving alternate high and low speed centrifugation and aggregation of the protein by adjusting the *pH* to 4.7. This yields a protein remarkably similar in composition to TMVP. The amino acid sequence of PM2P is identical with that shown in Fig. 1 of Chap. 3 for TMVP, except that isoleucine replaces threonine at position 28 and aspartic acid replaces glutamic acid at position 95, (20, 21). These two amino acid replacements result in a protein which is incapable of coating its own RNA and which exhibits unusual polymerization properties. SIEGEL, HILLS and MARKAHM (22) found that in 0.1 μ ammonium acetate buffer at *pH*=5.2 single, double and double-double open helicies were obtained. This result has been confirmed by SRINIVASAN and LAUFFER (18), as shown in Fig. 13a and b.

In an attempt to investigate the possible relationship of these open helicies to the normal helicies found as the end result of late stage polymerization of TMVP, SRINIVASAN and LAUFFER (18) carried out experiments comparable to those of SMITH and LAUFFER (23) on TMVP, already described in Chap. 4. In

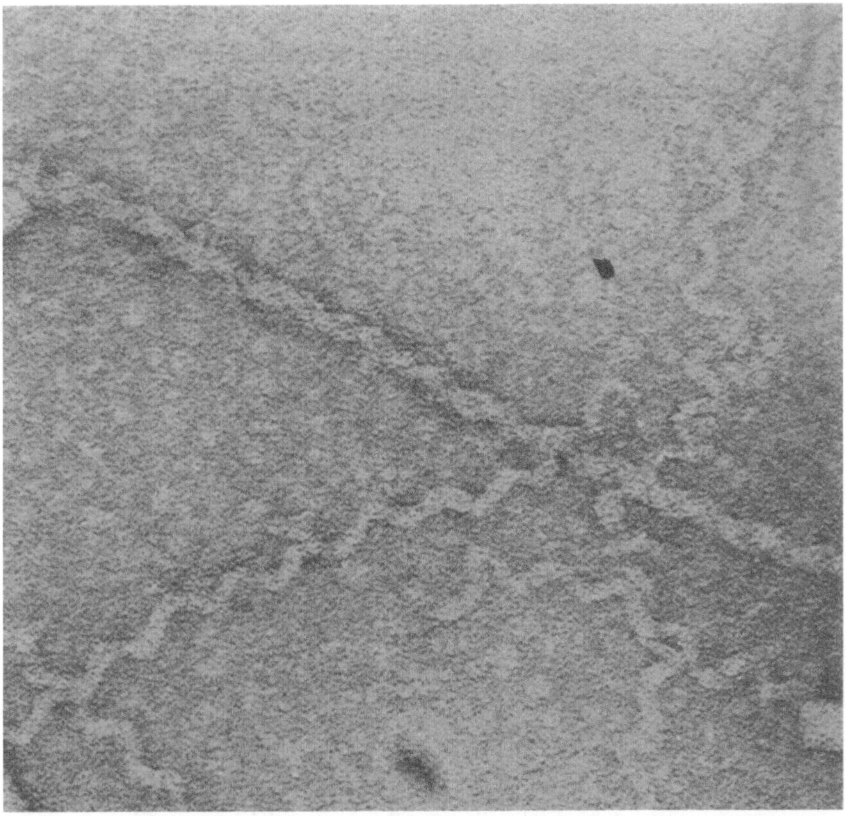

(a)

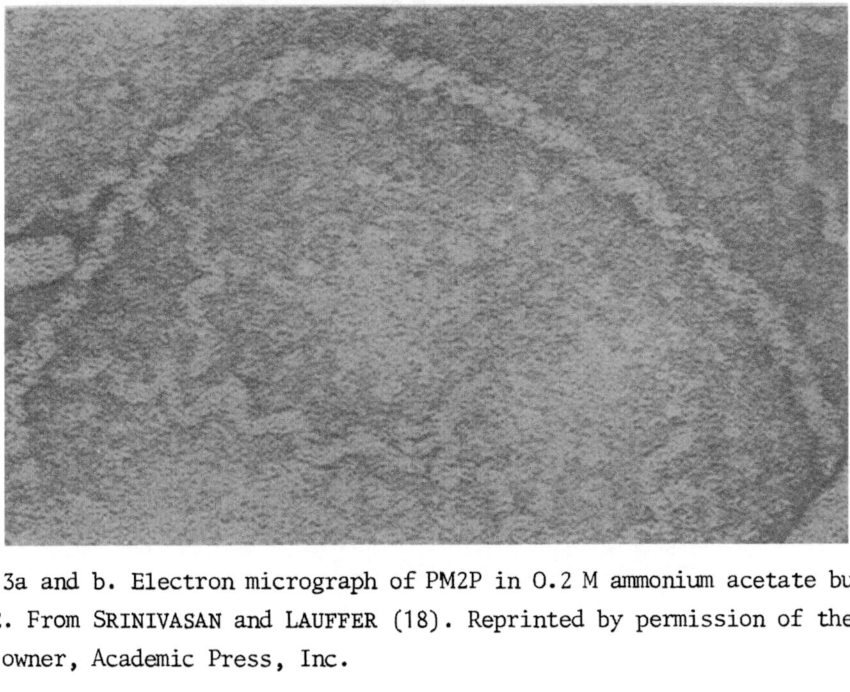

(b)

Fig. 13a and b. Electron micrograph of PM2P in 0.2 M ammonium acetate buffer, *pH*=5.2. From SRINIVASAN and LAUFFER (18). Reprinted by permission of the copyright owner, Academic Press, Inc.

0.1 μ ammonium acetate buffer at *pH*=5.2, the protein failed to polymerize normally but denatured when the temperature was raised to room temperature. However, in 0.1 μ phosphate buffer at *pH*=5.5, results shown in Fig. 14, almost identical to those with TMVP at *pH*=6.5 were obtained. Electron micrographs prepared at 4 and 18°C showed the results exhibited in Fig. 15 and 16. At 4° there were twisted ropes definitely thinner than polymerized TMVP rods. However, on a sample stored at 4°C for three months, rods of the normal thickness were observed. At 18° there were double discs and long rods of normal thickness. It is apparent that PM2P can polymerize to form double discs, stacked

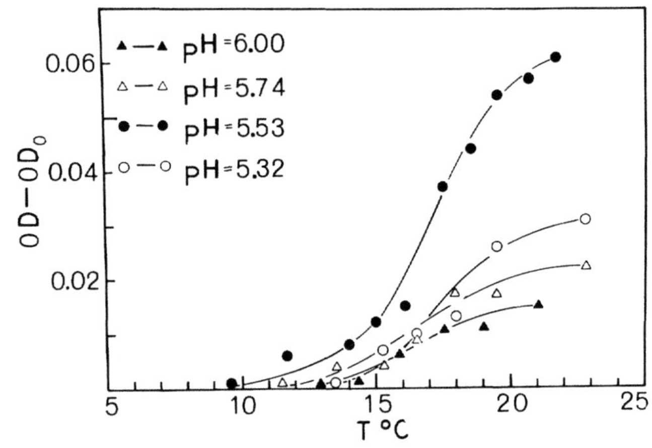

Fig. 14. Optical density at 320 mμ vs. temperature. Effect of *pH* on polymerization of PM2P in 0.1 μ phosphate buffer. From SRINIVASAN and LAUFFER (18). Reprinted by permission of the copyright owner, Academic Press, Inc.

double discs, normal helical rods and abnormal or thin helical rods. The most remarkable difference between the polymerization of PM2P and TMVP is that it is necessary to drop the *pH* to 5.5 in order to obtain the type of polymerization with increasing temperature observed for TMVP at *pH*=6.5.

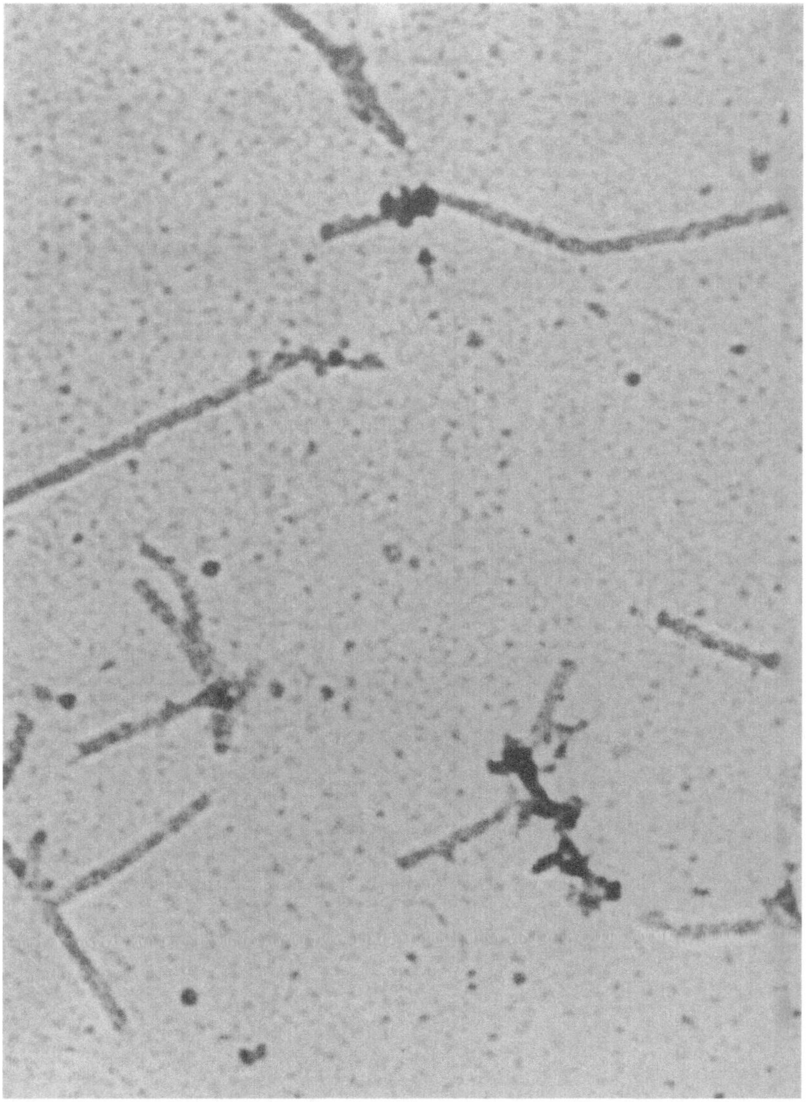

Fig. 15. Electron micrograph of PM2P in 0.1 ionic strength, phosphate buffer, *pH*=5.53; *T* = 4°C; width of particles = 100-120 Å. From SRINVASAN and LAUFFER (18). Reprinted by permission of the copyright owner, Academic Press, Inc.

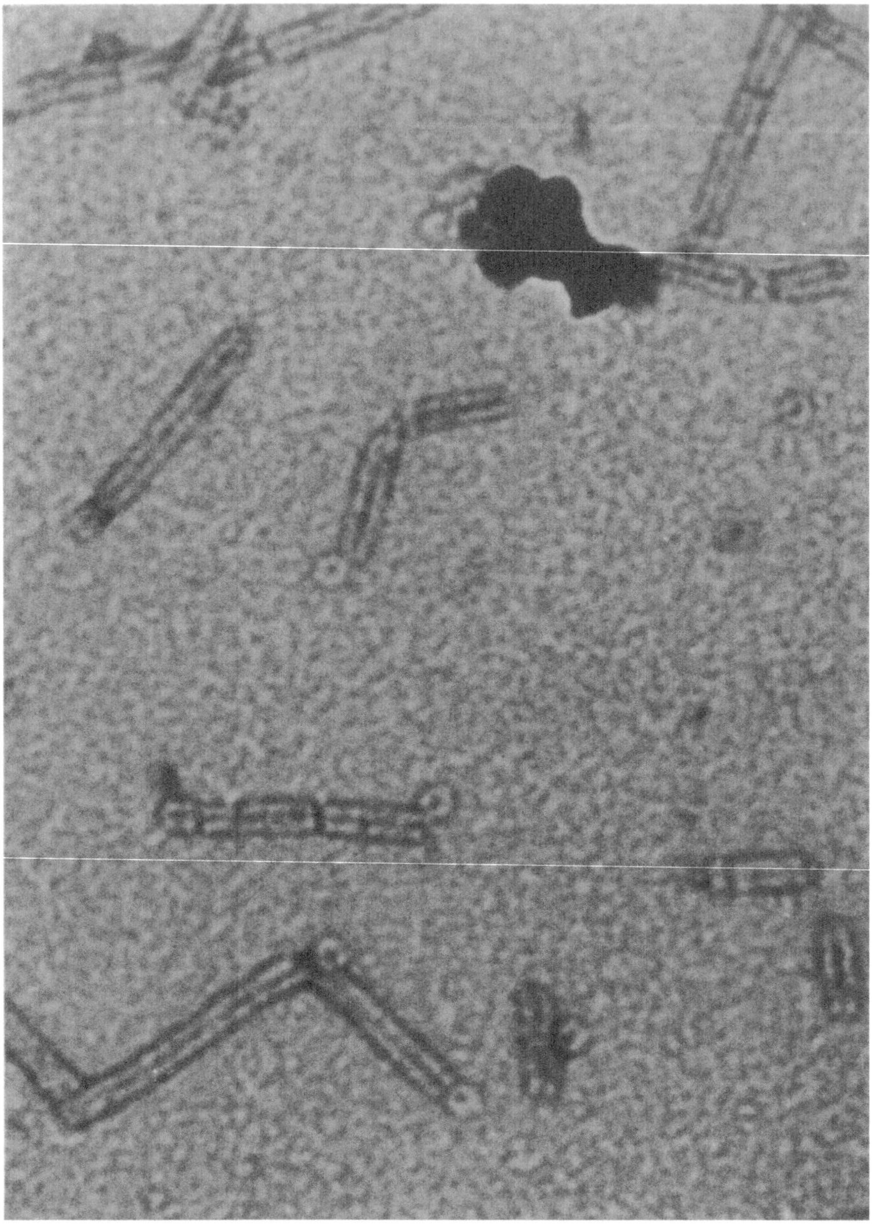

Fig. 16. Electron micrograph of PM2P in 0.1 ionic strength, phosphate buffer, pH=5.53; T = 18°C; width of particles = 180 Å. From SRINIVASAN and LAUFFER (18). Reprinted by permission of the copyright owner, Academic Press, Inc.

Cucumber Green Mottle Mosaic Virus Protein

In studies designed primarily to investigate reconstitution, OKADA *et al.*, (24) and OHNO *et al.*, (25) provided some information about the polymerization of the protein from the watermellon strain of cucumber green mottle mosaic virus (CGMMV-W). This virus has the general appearance of TMV, a rod of the same length and width. Because of its chemical and immunological properties, CGMMV-W is considered to be somewhat related to TMV (25). Its protein can be assembled with its own RNA or with TMV RNA.

Under conditions of low ionic strength (*pH*=7.2 pyrophosphate buffer at concentrations of 0.05 M or lower) or at temperatures below 10°C, the protein exists primarily in the form of units with sedimentation coefficients of 2.5 S. This value is intermediate between the 1.9 S of TMV protein subunits in the fully dissociated form and the 3 S of cucumber mosaic virus protein, assumed to represent dimers. It could represent a system containing monomers and dimers in approximately equal concentrations. The protein aggregates when temperature or ionic strength are raised (24). In 0.1 M pyrophosphate buffer at *pH*=7.2 at 25°C, the dominant form is a component with a sedimantation coefficient of 13 S. Calculations comparable to those which show that a sedimentation coefficient of approximately 20 S corresponds to a TMVP double disc indicate that 13 S corresponds to a single disc. As will be shown in Chap. 6, 13 S material is necessary to initiate reconstitution of this protein with RNA, but not for the elongation of reconstituted rods once initiation has been achieved. From the fact that there is a lag of about 45 minutes when protein in the unaggregated form, 2.5 S, is added to RNA but no lag when protein in the disc form, 13 S, is used, it was concluded that the formation of single discs is a slow process, somewhat resembling the formation of double discs of TMVP. It is probably significant that the apparent absence of trimers is correlated with the formation of single rather than double discs.

Tobacco Rattle Virus

The etiological agent of tobacco rattle is compound. Two types of rod-like virion, each crudely similar to TMV, are required to produce typical infections. Long RNA rods, about 190 mμ, code for RNA replication and short rods, varying between 50 and 115 mμ, code for protein synthesis (26). The protein coats of the several particles seem to be identical; the difference resides in the RNA's.

Coat protein can be extracted and purified by methods comparable to those used for the purification of TMV protein. The polymerization properties of tobacco rattle virus protein, TRVP, from two strains of the virus have been investigated by MORRIS and SEMANCIK (27) and FRITSCH *et al.*, (28). MORRIS and SEMENCIK describe three classes of protein aggregate for the *C* strain. The smallest have sedimentation coefficients of 10 Sverdbergs or less. The state of aggregation of this material was not investigated in detail; it is called 10 S material. A distinct aggregate sedimented as a single sharp boundary in the analytical ultracentrifuge with a value of s_w^{20} of 37. Electron micrographs indicated that this material consists of double discs somewhat like those of TMVP, except that they are larger. MORRIS and SEMENCIK calculated from the sedimentation coefficient of 37 S that there are 62 protein subunits of molecular weight 1900 per double disc. This particular component is called 35 S. In addition, materials with sedimentation coefficients of 50 S and higher were observed. These materials are called 50 S components.

At 11°C in 0.1 ionic strength buffers, 10 S material predominates at *pH*=4. At *pH* values between 4 and 9.5, 10 S and 35 S material coexist, with a maximum amount of 35 S at *pH*=7. At 21°C, 10 S, 35 S and 50 S material coexist in 0.25 M lysine buffers. At *pH*=4 most of the protein is in the 10 S, an appreciable amount in the 50 S, and very little in the 35 S form. At *pH* values above 4, less material is in the 10 S fraction and more in 35 and 50 S. The maximum amount of 35 S was observed near *pH*=7 and the maximum amount of 50 S at about *pH*=8. Somewhat similar results were obtained by FRITSCH *et al.* (28) with protein from the CAM strain of the virus. At 2°C, they observed components with sedimentation coefficients of 2-4 S and 28-36 S. They calculated that 2 S corresponds to protein monomer and 4 S to the dimer. The 36 S component corresponds to double discs; the 28 S component was not identified. As in the results with the *C* strain, raising *pH* and lowering ionic strength promote the formation of double discs. The effect of temperature is less pronounced than with the *C* strain; nevertheless, raising the temperature to 30°C resulted in the formation at high *pH* values of small amounts of material sedimenting between 50 S and 100 S. Electron micrographs show mainly double discs and short stacks of discs.

The effect of the molarity of phosphate buffer at *pH*=6.5 and 4°C on the conversion of 10 S material of the *C* strain to 35 S material was investigated. At a molarity of .001, the material is entirely in the 35 S state. As molarity is increased to 0.5, 10 S material increases and 35 S decreases to almost zero at the highest buffer concentration (27).

The conversion of discs into larger aggregates was studied as a function of temperature and buffer molarity at *pH*=6.5 by the method of light scatter (27). The results are shown in Fig. 17. The polymerization is clearly endothermic, resembling rod-forming polymerization of TMVP as far as the effect of temperature goes. However, quite opposite to the TMVP case, the extent of polymerization is inversely related to ionic strength.

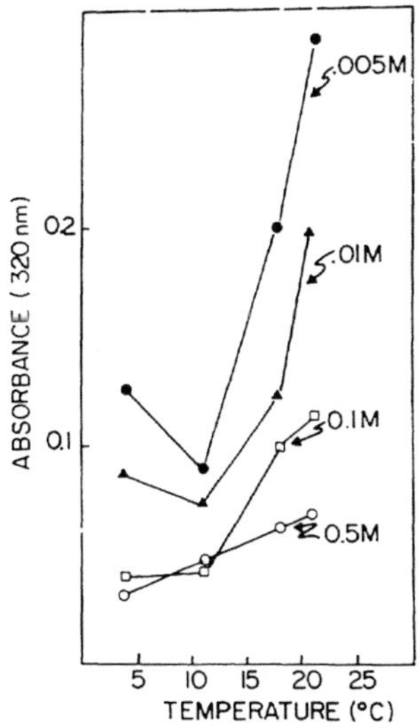

Fig. 17. The effect of buffer molarity on temperature induced polymerization of TRV protein (1 mg/ml) into large aggregates as determined by light scattering at 320 nm. The protein was dialyzed against KPO_4 buffers of 0.005-0.5 M, and the temperature was slowly increased over a 24 hr period. From MORRIS and SEMANCIK (27). Reprinted by permission of the copyright owner, Academic Press, Inc.

SEMANCIK (29) reported that TRVP has twice as many basic amino acid residues and roughly the same number of dicarboxylic residues per protein subunit as TMVP. Unfortunately, amides and free acidic groups were not differtiated. This might mean that the protein has an isoelectric point near neutrality. The protein does have a positive charge at *pH*=4.5 (29). If the isoelectric point is that high, it might account for the fact that polymerization is optimal at neutrality or slightly higher *pH* values. The ionic strength effect on polymerization of TMVP can be explained in terms of "salting-out". By analogy, the opposite effect of ionic strength on polymerization of TRVP would be explained in terms of "salting-in". Highly polar proteins, such as globulins, do exhibit the "salting-in" phenomenon. The amino acid composition of TRVP (29) does indicate that it is more polar than TMVP.

Speculation on Models for Polymerization of Coat Protein of TMV and Its Strains

Two main lines of thought have dominated the attempt to understand the facts uncovered about rod-forming polymerization of the protein from TMV and its strains. As early as 1962, LAUFFER (30) focused attention on the fact that, during the early stages, polymerization follows the mathematics of condensation polymerization, as discussed in detail in Chap. 4. In contrast, CASPAR focused attention on intermediates in the polymerization process (17). He assumed the existence of six bonding sites on each TMVP monomer, located at the points of contact between monomeric particles in the hexagonal net formed when they are fully polymerized. He postulated a number of stable intermediates; the cyclical trimer, a cyclical heptamer with one monomeric unit in the middle surrounded by six others, the double disc and a three-ringed disc consisting of 48 monomers. He conceived of the ultimate polymerization process as consisting of a series of steps going from one intermediate to another. These two points of view are not necessarily incompatible. There is agreement that at least some of the intermediates postulated by CASPAR do in fact exist. The evidence for double discs and stacked double discs is compelling. BANERJEE and LAUFFER (31) showed that at temperatures below $10^{o}C$, osmometrically determined number average molecular weights were in the neighborhood of 50,000 at very low protein concentrations. This is at least consistent with the idea that a cyclical trimer, with molecular weight 52,500, is a stable particle.

The real problem has to do with the interpretation of experimental data such as those shown in Fig. 9 of this chapter and Fig. 1 of Chap. 4. Such curves fit the mathematics of condensation polymerization at intermediate temperatures. But what units are polymerizing to give the increases in optical density or turbidity? Initially, LAUFFER (30) assumed that the polymerizing unit was an intermediate with a molecular weight of approximately 100,000, based on the diffusion and sedimentation results of SCHRAMM and ZILLIG (32). SMITH and LAUFFER (23), however, consistent with the finding of BANERJEE and LAUFFER (31), assumed that the polymerizing unit involved in rod-forming polymerization was the cyclical trimer. Nevertheless, they did point out that the selection of this as the polymerizing unit was arbitrary and that the mathematical analysis is not greatly affected by postulating particles of higher molecular weight as the polymerizing unit.

As was discussed in detail in Chap. 4, for a number of reasons, including the finding of SRINIVASAN and LAUFFER (13) that DTMVP at low temperatures in 0.1 μ phosphate buffer at *pH*=6.25 exists in the form of particles the size of double discs or even somewhat larger, the author has adopted the point of view that light scatter measurements are sensitive primarily to the further polymerization of particles the size of double discs, probably the hypothetical double spirals. Whatever the final resolution of this issue may be, it is in one sense not critically important. It is obvious that the free energy of polymerization varies sharply with temperature between positive values at low temperatures and negative values at somewhat higher temperatures. Thus, particularly below the middle of the temperature range studied, the value of the negative free energy is small. It is undoubtedly the algebraic sum of a number of terms, positive and negative. Several of these contributions have been identified. The sole negative contribution for which we have detailed evidence, presented in Chap. 2, stems from the increase in entropy associated with release of water molecules upon polymerization (33, 34). There must be some additional negative contribution; polymerized rods are stable even in the dry state, where water - interaction terms are necessarily absent. The positive free energy contribution involved in the polymerization process stems from the binding of hydrogen ions (35), from changes in rotation and translational components of the entropy as estimated with the Sackur-Tetrode Eq. (36), and from the electrical work associated with polymerizing charged cylinders (37). One can summarize with Eq. (1).

$$\underset{\substack{\text{net}\\ + \text{ or } -}}{\Delta F^{\circ}} = \underset{\substack{\text{water}\\ \text{release}\\ -}}{\Delta F^{\circ}} + \underset{\substack{\text{Sackur-}\\ \text{Tetrode}\\ +}}{\Delta F^{\circ}} + \underset{\substack{\text{el}\\ +}}{\Delta W} + \underset{\substack{\text{H}^{+}\\ \text{binding}\\ +}}{\Delta F^{\circ}} + \underset{\substack{\text{unknown}\\ + \text{ or } -}}{\Delta F^{\circ}} \qquad (1)$$

Let us analyze the polymerization of TMVP in 0.1 μ phosphate buffer at *pH*=6.0 on the assumption that the polymerizing unit is the size of a double disc consisting of 34 protein monomers. SRINIVASAN and LAUFFER (13) recalculated the polymerization data for these conditions to indicate that ΔH° = +371 kcal/mole and ΔS° = +1357 e.u. With the aid of Eq. (1) of Chap. 1, one can calculate that at 290°K, ΔF° is -23 kcal/mole. This corresponds to $\Delta F^{\circ}_{\text{net}}$ in Eq. (1). The entropy contribution arising from changes in translational and rotational freedom upon polymerization was calculated by LAUFFER (36) to be approximately -100 entropy units. At 290°K, this provides a contribution to the free energy of +29 kcal/mole, corresponding to $\Delta F/_{\text{Sackur-Tetrode}}$ in Eq. (1).

As was shown in Chap. 4, based on the treatment published by ANSEVIN, *et al.* (35), for the binding of 1 mole H^{+} by typical carboxyl groups on proteins dissolved in phosphate buffer at *pH*=6, ΔS° is approximately -6 e.u. and ΔH° is -123 cal/mole. Since SCHEELE and LAUFFER (15) showed that at *pH*=6.0 about one mole of hydrogen ion is bound per protein monomer upon polymerization, and since there are 34 monomers in a double disc, the contribution to ΔH° from hydrogen ion binding should be -4.3 kcal/mole of double disc polymerized and ΔS° should be -204 e.u. The contribution to ΔF° at 290°K is, therefore, +55 kcal/mole. However, a note of caution must be introduced. If, as will be discussed later, double spirals with the same molecular weight as double discs are the polymerizing unit in rod-forming polymerization and if some of the hydrogen ions are bound during the formation of the double spirals from smaller units, then the number of hydrogen ions bound during further polymerization will be only a fraction of the total here considered, and the positive free energy change of this process will be less than 55 kcal/mole.

With the aid of Eq. (18) of Chap. 4, it is possible to calculate ΔW_{el}. For the reaction, $2P \rightarrow P_2$, ΔW_{el} equals W_{el} for P_2 minus $2W_{el}$ for P. With appropriate values of the parameters for a 0.1 μ solution at 25°C, the values shown in Table 2 of Chap. 4 are obtained for ΔW_{el} in kcal/mole for dimerization of TMVP rods of various lengths and various electric charge per protein monomer. It is also possible to calculate ΔW_{el} for a reaction involving the addition of one double spiral to the end of a very long rod. It turns out that ΔW_{el} for this process is equal to +8.07 z^2, where z is the charge per protein monomer.

SCHEELE and LAUFFER (15) have shown that at pH=6.0, z has a value of -2. Thus, ΔW_{el} has a value of +32 kcal/mole. Since this term is purely one of enthalpy, ΔH is also +32 kcal/mole for this contribution to the total process.

From Eq. (1), it is thus apparent that $\Delta F^o_{water\ release} + \Delta F^o_{unknown}$ must equal approximately -139 kcal/mole. The question is wether or not this is a reasonable figure.

Since there is no information concerning $\Delta F^o_{unknown}$, the only way to deal with the problem is to assume that this is small and attribute the entire term to water release. STEVENS and LAUFFER (33) showed that 0.027 g of water were released per gram of protein upon polymerization. This corresponds to 16,100g of water per mole of double spiral. The problem now is to guess the free energy decrease associated with the release of this amount of water. First, one observes that ΔS^o for this process must be 1661 entropy units, in order to overbalance the negative contributions of 100 and 204 and leave a net of 1357. Similarly, ΔH^o attributed to this process must be 343 kcal. Therefore, it is reasonable to assume that the free energy for water release would be zero at at temperature equal to 343,000/1661 or 206°K. The free energy at 290°K would then equal -(290 - 206) ΔS^o, or -84 ΔS^o.

Therefore, a necessary step in the process is to obtain some estimate of the entropy increase to be expected when one gram of water is released by protein. One possible estimate can be obtained by considering the entropy increase when one gram of ice melts at 0°C. This is +0.292 kcal/g. deg. This is probably too high an estimate, because water molecules are very highly oriented in ice. Another possible estimate can be obtained from the data of BULL (38), who studied the water taken up by dry proteins in equilibrium with water vapor. BULL measured water uptake by the proteins listed in Table 1, by measuring the weight of protein plus water in equilibrium with relative vapor pressures of water ranging from 0.05 to 0.95. He recorded the number of grams, a_i, of water in equilibrium with each relative pressure, x_i, taken up by 100 g of each protein.

In a reversible isothermal process involving no non-mechanical work, $dF = VdP$. For an ideal gas, $PV = nRT$ and $dF = RTnd\ln P$. When n_i moles of water are vaporized from pure water at its equilibrium vapor pressure, P_o, reduced isothermally and reversibly to pressure P_i and finally bound to protein in equilibrium with vapor at P_i,

$$\Delta F_i = 0 + RTn_i \int_{P_o}^{P_i} d\ln P + 0 = RT\, n_i \ln\frac{P_i}{P_o} \equiv RTn_i \ln x_i \qquad (2)$$

The total free energy change for binding n_s moles of water vapor, ΔF, is $\Sigma\Delta F_i$

$$\Delta F = \Sigma\Delta F_i = RT\Sigma\, n_i \ln x_i = RT\Sigma \ln x_i^{n_i} = RT\int_1^0 d\ln x_i^{n_i} = RT\int_1^0 \frac{d(x_i^{n_i})}{x_i^{n_i}}$$

$$= RT\int_1^0 \frac{n_i x_i^{n_i - 1} dx_i}{x_i^{n_i}} = RT\int_1^0 \frac{n_i}{x_i}\, dx_i = -RT\int_0^1 \frac{n_i}{x_i}\, dx_i = \frac{-RT}{M}\int_0^1 \frac{a_i}{x_i}\, dx_i \qquad (3)$$

where a_i is the number of grams of water bound at relative pressure, x_i , and M is the molecular weight of water.

Table 1

	ΔF_{25} cal	ΔH cal	ΔS cal/deg.	a_s g	$\Delta S/a_s$ cal/g·deg
silk	-656	-1270	-2.07	33	-0.061
wool	-975	-1950	-3.27	41	-0.080
B-zein	-605	-1160	-1.86	29	-0.064
C-zein	-604	-1060	-1.53	29	-0.053
elastin	-1000	-2250	-4.20	47	-0.089
collagen	-1620	-3640	-6.77	62	-0.110
gelatin	-1537	-3800	-7.60	85	-0.090
egg alb. (lyophylized)	-892	-1570	-2.27	47	-0.048
egg alb. (unlyoph.)	-956	-1350	-1.32	51	-0.026
egg alb. (denatured)	-814	-1400	-3.36	36	-0.093
β-lactoglob. (lyoph.)	-917	-1610	-2.32	53	-0.044
Serum alb.	-1034	-1450	-1.40	46	-0.031
α and β-pseudoglob.	-1109	-1830	-2.42	58	-0.042
γ-pseudoglob.	-1119	-1700	-1.95	63	-0.031

Thermodynamic parameters for binding of water by 100 g of various proteins. Data of BULL (38).

BULL's data in their original form show a_i as a function of x_i for each protein. The data extrapolate to a value of zero for a_i when x_i is zero, and on the other end of the scale they extrapolate to the saturation value of a, a_s, when x_i is equal to one. The integral in Eq. (3) was evaluated by plotting a_i/x_i against x_i and measuring the area under the curve between $x_i = 0$ and $x_i = 1$. In this way, the values of ΔF_{25} listed in Table 1 were obtained. The measurements were repeated at 40°C; with the aid of Eq. (1) of Chap. 1, ΔH and ΔS were calculated for each protein, on the assumption that ΔH and ΔS remain constant over the temperature interval studied. BULL very properly cautioned the reader concerning the dangers of this assumption and the inherent inaccuracy in the measurements. Nevertheless, the values listed in Table 1 for ΔH and ΔS are his best available estimates.

The values for ΔF_{25}, ΔH and ΔS shown in Table 1 are for the process of binding a_s grams of water to 100 g of protein. The values of $\Delta S/a_s$ in the extreme right hand column are thus estimates of the entropy changes in calories per degree per gram of water bound. These entropy changes range from -0.026 to -0.11. The entropy changes for releasing water would have the same values but positive signs.

It must be emphasized that the water binding studied by BULL cannot be exactly the same as that involved in the present study. In BULL's case, the free energy of binding is negative at room temperature; we are considering a hypothetical process in which the free energy of water binding is positive at room temperature. It is, nevertheless, not inappropriate to assume that the degree of structure found in the bound water as studied by BULL is comparable to the degree of structure in our case. The difference could be largely in ΔH. Based on BULL's results, we will assume that the entropy increase when water is dissociated from protein is somewhere between + 0.026 and + 0.11 cal per gram degree. The entropy values thus calculated are average values for all water bound by each protein. MCLAREN and ROWEN (39) calculated entropy as a function of the amount of water bound from Bull's data; ΔS for the trasfer of one gram of water from gelatin, silk or wool to liquid water varies between +0.05 and +0.1 cal per degree for all except the most tightly bound water molecules, for which the values are much higher.

ΔS for release of 16,100 g of water should have some value between +418 and +1770 eu. if the entropy change in the process under consideration resembles the averages calculated from Bull's data. ΔF at 290°K is calculated to be

-35 kcal/mole when the lower figure is used and -149 when higher is used. The latter is close to the value mentioned above, -139 kcal/mole needed to balance Eq. (1). If, as was explained above, the estimate of the positive free energy contribution for hydrogen ion binding in the polymerization of double spirals is too high, then the estimate of the negative free energy contribution from water release, -139 kcal/mole, is also too high. A possible value is something near -100. This is still within the range -35 to -149. This line of reasoning is too highly speculative to prove anything positive. The only thing one can conclude with safety is that the idea that the major contribution to the negative free energy comes from the known amount of water released cannot be discarded on the basis of the evidence considered.

It is appropriate at this point to consider in greater detail the scheme for the polymerization of the proteins from TMV and several of its strains introduced in Chap. 3. For convenience, it is restated below. The symbol, A, means relatively unpolymerized protein such as that found at $4^{\circ}C$ in a 1 mg/ml solution in 0.1 μ phosphate buffer at pH=6.5 with a number average molecular weight of approximately 50,000. The symbol, R, means the short rod or double spiral, D means double discs such as those described in Chap. 3, P means rod-like polymers in which the protein subunits are arranged helically as they are in the virus itself, and S means stacked double discs. The symbols, n, m, x, y and z are merely the numbers required to make the equations balance.[1] Solid arrows represent experimentally established reactions, and dotted arrows correspond to possible reaction.

---

[1]It is possible to speculate about the numerical values of x, y, and z. The fact that two hydrogen ions are bound per protein subunit in going from pH=7.5, where the material is largely in the state of trimers, to about pH=5.8, where it is highly polymerized (15), since indications are that no hydrogen ions are bound during the formation of trimers, suggests that $(x+y)$ is 2 per subunit. If z is 1 per subunit (see page 67) then $(x+y-z)$ is 1 per subunit. It seems that z represents hydrogen ions bound when subunit contacts around the turn of a disc or of a helix are made. Both y and $(x-z)$ would then represent hydrogen ions bound in making contact between neighboring turns of the helix. Since half of such contacts should be made in the formation of a double spiral and the other half in the polymerization of double spirals, both y and $(x-z)$ can be assigned a value of 1/2 per subunit. It follows that x would be 3/2 per subunit.

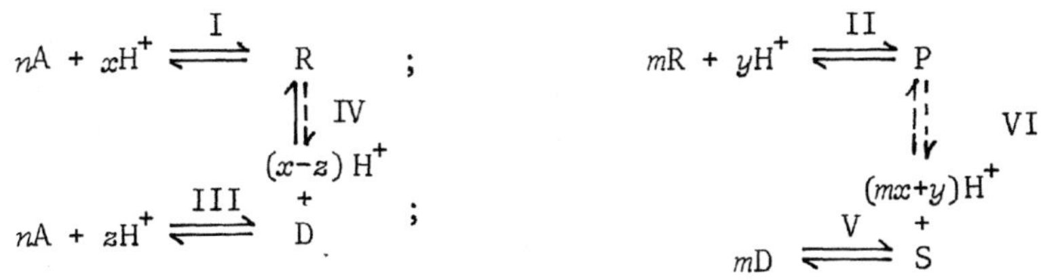

This scheme preserves some of the features of earlier models and rejects others. It allows for the possibility that double discs can be intermediates in the polymerization process, as first suggested by CASPAR (17) but rejects the idea that they are necessary intermediates. It preserves the key idea in the model first proposed by the author (36, 41) to the effect that alternative pathways are available for polymerization of particles even as small as molecular weight 50,000. There must be alternative pathways, not only because one type of polymerization is highly *pH* - sensitive while the other is relatively insensitive to *pH*, but also because of new evidence presented by DURHAM and KLUG (42). They found that reaction III is very slow, requiring approximately one day to proceed to half completion. In contrast, the polymerization studies of ANSEVIN and LAUFFER (43) showed that under a wide variety of experimental conditions, polymerization in 0.1 μ phosphate buffer to form helical rods is nearly complete in a few minutes. The model presented above rejects the hypothesis advanced in a variant (44) of the author's model, a hypothesis that process I leads to the formation of thin open helicies like those obtained when PM2P polymerizes in ammonium acetate buffer, as shown in Fig. 13. The reason for rejecting this assumption is that extensive efforts to find such intermediates in strains other than PM2 have been largely unsuccessful. A further reason is the innate attractiveness of the hypothesis of DURHAM *et al.* (40) that double discs can be converted by lowering the *pH* into "lock washers" or, as modified by the author, into double spirals. This feature is incorporated in the model as reaction IV. In the model presented above, reactions I and III are in competition. When the hydrogen ion concentration is increased, reaction I will have a competitive advantage over III.

The binding of hydrogen ions in process I at *pH* values in the neighborhood of 6 is a thermodynamic liability. Hydrogen ions are bound only because, concomitant to polymerization, negative groups on the surface are put in a position in which hydrogen ion binding is a necessity. This could result from a negative group being removed from contact with water and placed in contact with an organic protein surface of low dielectric constant, as suggested by ANSEVIN *et al.* (35), or from a negative group being brought into close proximity with another negative group, as suggested by CASPAR (17). While both hypotheses can explain hydrogen ion binding, the latter affords an attractive explanation for the fact that divalent cations such as lead and calcium are strongly bound by TMV but not by unpolymerized TMVP. Nevertheless, the former cannot be ruled out on this basis, because many lead and calcium soaps are more soluble in organic solvents such as ether and alcohol than in water. The two hypotheses are not radically different; both require groups which are normally ionized at *pH*=6 or 6.5 to be at least partially buried within the crevices between adjacent protein subunits. In either case, hydrogen ions will be bound at *pH* values at which carboxyl groups are normally ionized, and there must be an external source of negative free energy to push the polymerization in spite of this thermodynamic liability. It is probable that the "burying" involves bringing carboxyl groups from adjacent monomers close together as well as reducing the average dielectric constant in the neighborhood of the groups partway toward the level of pure organic material. Because of the fact that process III involves obligatory binding of fewer hydrogen ions, one is forced to conclude that a different kind of bonding exists in this case, a kind which involves bringing negative ions neither too close together nor into contact with an organic surface. In the author's original model, this necessity was recognized and was attributed to different rotations of the monomeric units about their long axes. However, differences in conformation within the subunits or differences in orientation of the long axes, such as parallel versus non-parallel, could also account for this. The important point is that the bonding in process I must be basically different from the bonding in process III, which means differences in the surfaces which come into contact when polymerization proceeds.

Reaction II is well established at *pH* values of 6.5 and below; rods with helical arrangements of protein subunits are formed when temperature is increased and are dissociated when temperature is lowered. The forward part of reaction V proceeds readily at high concentrations, high *pH* values and high ionic strength. Stacked double discs are formed readily under these conditions. DURHAM and KLUG (42) report that sometimes the stacked double discs are difficult to dissociate, from which one must conclude that sometimes the backward portion of reaction V proceeds with difficulty. Reaction VI is speculative.

The essential features of this proposed scheme are consistent with data of SCHUSTER and SCHEELE (45). They carried out experiments in which temperature was raised suddenly from 18° to 19°C. Such a temperature increase results in an increase in polymerization, which can be followed optically as an increase in turbidity. Since polymerization is accompanied by hydrogen ion binding, in unbuffered solutions there is a slight increase in *pH* when the temperature is increased. This increase in *pH* was followed by changes in the absorbance of *pH*-indicator dye, chlorophenol red. SCHUSTER and SCHEELE observed up to three relaxations when temperature was suddenly increased. The slowest of the three reactions had a reciprocal relaxation time of less than one sec^{-1}, whereas the two faster reciprocal relaxation times fell in the range between 2 and 70 sec^{-1}. In the case of the two faster reactions, it was possible also to measure the amplitudes of the changes, which are measures of the extent of change. These amplitudes are shown as a function of *pH* for the change in degree of polymerization in Fig. 18 and for the change in hydrogen ion binding in Fig. 19.

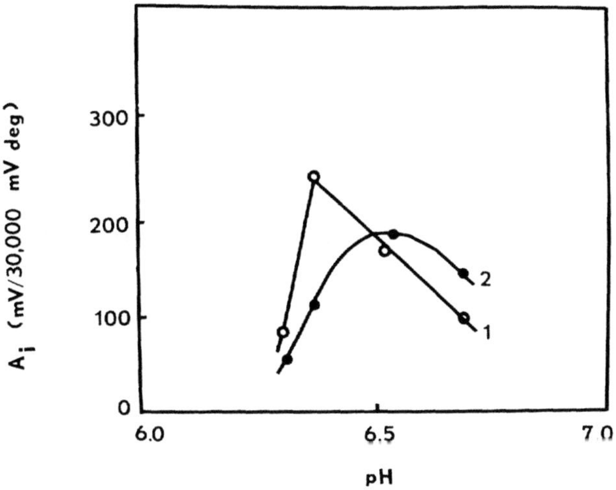

Fig. 18. *pH* dependence of the turbidity change when the temperature of TMVP at 5.0 mg/ml in 0.10 μ phosphate buffer is raised suddenly from 18° to 19°C. Open circles represent the fastest reaction, closed circles the intermediate. Data of SCHUSTER and SCHEELE (45)

The polymerization data of Fig. 18 are in general accord with the findings from the author's laboratory. At *pH* values a few tenths below 6.5, there is little further polymerization when temperature is increased from 18 to 19°C because the protein is polymerized almost to its saturation level at 18°. There is also very little polymerization at *pH* values a few tenths above 6.5. Here the

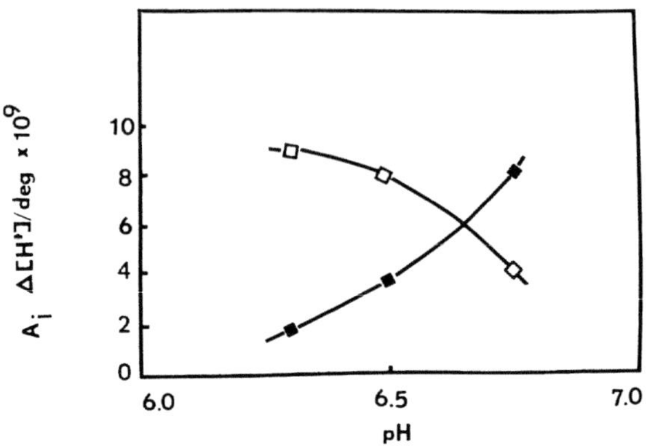

Fig. 19. *pH* dependence of the extent of hydrogen ion binding when the temperature of TMVP at 5.0 mg/ml in 0.10 M KCl and 1.7 x 10^{-5} M chlorophenol red is raised suddenly from 18^{0} to 19^{0}C. Open squares represent the fastest process, closed squares the intermediate. Data of SCHUSTER and SCHEELE (45)

situation is just the opposite; protein polymerizes only to a small extent even at the higher temperature. In contrast, at *pH*=6.5 there is only partial polymerization at 18^{0} and significantly greater polymerization at 19^{0}.

The increase in *pH* accompanying a temperature jump of one degree from 18^{0}C, shown in Fig. 19, is more instructive. It must be observed, however, that the extent of hydrogen ion binding indicated by these results is remarkably small. The total change in hydrogen ion concentration by both processes is about 10^{-8} mole/l. Protein concentration calculated on the basis of monomer is about 3 x 10^{-4} mole/l. In the overall polymerization near *pH*=6.5, approximately one hydrogen ion is bound per monomeric unit.

SCHUSTER and SCHEELE have identified the fastest of the three processes with reaction II. The intermediate process was identified with reaction IV and the slowest process with reaction III. SCHUSTER and SCHEELE did not postulate reaction I; it might also be involved in the slowest binding reaction, not capable of analysis by their method. There is no difficulty at all in interpreting the intermediate process in terms of this identification. At the lower *pH* value, the extent of binding hydrogen ions at the intermediate rate is low because very little double disc is present; reaction III is at a competitive disadvantage with reaction I. As *pH* is increased, more and more double disc is available because reaction I loses its competitive advantage. At *pH*=6.5 the double spirals, formed from double discs, polymerize further, but the rate limiting step is reaction IV. At *pH*=6.75, the double spirals formed do not undergo further polymerization to a great extent because of the scarcity of hydrogen ions required for reaction II.

To understand the fastest hydrogen ion binding, however, it is necessary to postulate that the binding of hydrogen ions by already polymerized protein is also a fast process akin to the binding of hydrogen ions in reaction II. That hydrogen ions are bound by polymerized protein when the *pH* is dropped slightly at constant temperature from *pH*=6.3 is shown by the titration results of SCHEELE and LAUFFER (15). The binding of hydrogen ions by the fast process at *pH*=6.3 probably comes about by "burying" negative groups as a result of a tightening up of the structure of the helical rods upon increase of temperature, comparable to that observed by calorimetric means (46), and by volume changes (47) when *pH* is lowered.

Support for this reaction scheme is also found in the work of WESTOVER (48), described in Chap. 4. WESTOVER found that, for TMVP in 0.1 μ phosphate buffer at 12°C, the "cut-off" concentration was 1.10 milligrams per milliliter at *pH*=6.5, and 0.10 mg/ml at *pH*=6.0. By differentiating Eq. (10) of Chap. 4 with respect to *pH* at constant temperature, treating α as a constant, one finds that $d \ln C_c/dpH$ equals $(1/RT)(d\Delta F^{o}/dpH)$. From the data this turns out to be +2.7 kcal/mole. The change in free energy with change in *pH* for binding hydrogen ions is an entropic effect. ΔS for binding decreases 4.56 entropy units per *pH* unit increase. At 285°K, this amounts to an increase in free energy of +1.3 kcal/mole of H^{+} ion bound. Thus, the change in "cut-off" concentration can be explained by the binding of approximately two hydrogen ions per mole of polymerizing unit in the reaction represented by I in the scheme. While the data obtained by WESTOVER are consistent with the idea that ΔH^{o} for polymerization is independent of *pH*, they are not sufficiently accurate to make it experimentally certain that the change in "cut-off" concentration with *pH* is purely an entropic effect. This weakens the argument somewhat.

Relation of Polymerization Behavior to Chemical Structure

The net free energy for rod-forming polymerization varies with *pH* and ionic strength in different ways for TMVP, DTMVP, PM2P and CV4P. It is instructive, therefore, to examine the known factors which contribute to the net free energy change. The positive term related to the change upon polymerization of the entropy of translation and rotation should be the same for the different proteins. Thus, this term probably does not contribute to the differences. Among the positive terms, there remain the electrical work for polymerization of charged rods and the free energy associated with binding of hydrogen ions which accompanies polymerization. It is necessary to examine both of these to see whether either or both will afford explanations for the differences in behavior.

As shown in the preceding section, an obvious parameter of the virus protein molecule which affects its ability to polymerize is its net charge. The proteins from all of the strains of TMV, TMVP, CV4P, DTMVP and PM2P fail to undergo rod-forming polymerization at *pH* values around neutrality, and all of them can be thus polymerized by merely lowering the *pH* to the neighborhood of 5.5. Electrophoretic mobility and hydrogen ion titration data on three of the proteins showed clearly that there is a reduction in the net negative charge per protein monomeric unit when *pH* is thus lowered. This is probably true of the fourth protein as well. Furthermore, treatment of TMVP and CV4P with succinic anhydride, which replaces positive charges with negative charges, has a drastic effect; it suppresses polymerization completely. One can conclude therefore that net negative charge on the protein particle opposes the process which causes polymerization, a process shown in Chap. 2 to be an entropic reaction involving interaction of the protein with water. It is obvious that when the negative entropic contribution remains the same but the positive electrical work contribution is sufficiently increased as a result of change in charge, the net free energy for polymerization can then become positive; polymerization will no longer occur.

Polymerization curves such as those shown in Figures 5 and 6 of this Chapter show that at *pH*=7 or 7.5 polymerization increases very gradually, if at all, when temperature is increased and the final state of polymerization is approximately what one would expect for double discs or short stacks of double discs. This is true for all of the proteins studied in detail. However, at some lower *pH* value for each protein, optical density and, therefore, degree of polymerization, increases sharply with temperature, and the maximum degree of polymerization obtained corresponds to relatively long rod-like particles. The approximate *pH* values at which this sharp change in polymerization pattern occurs is, in 0.1 μ phosphate buffer, 6.5 for TMVP, 6.25 for DTMVP and 5.5 for PM2P (Fig. 14). It is presumably approximately the same for CV4P as for TMVP. As shown in Fig. 7, the *pH* of transition drops to approximately 6 for DTMVP when the ionic strength is reduced from 0.1 to 0.01. The change brought about by reducing ionic strength with DTMVP can be understood in terms of the electrical work theory. Also, the difference between the behavior of TMVP and DTMVP at *pH*=6.5 can be explained by this theory if it is assumed that the negative entropic contribution to the free energy of polymerization derived from release of water molecules is approximately the same for the two proteins. DTMVP has two arginine groups fewer than TMVP and one less carboxyl group per protein subunit. Therefore, its net negative charge should be greater. Furthermore, electrophoretic mobility data on the polymerized proteins at *pH*=6.5, presented in Fig. 7, clearly show that DTMVP has a higher net negative charge at a given *pH* than does TMVP. Thus, one would expect that it would be necessary to suppress charge to a greater extent on DTMVP than on TMVP in order to permit long rods to form.

Of further interest are the experimental results in Fig. 3 and in Fig. 8 of Chap. 4, which show that more hydrogen ions are bound during temperature-mediated polymerization of DTMVP than of TMVP. As shown in Chap. 4, ΔH^o and ΔS^o are -123 cal/mole and -11.7 e.u., respectively, for the binding of a hydrogen ion by a protein COO^- in phosphate buffer (35). At *pH*=6.5, ΔS = -8.35 e.u. and at *pH*=6.25, ΔS = -7.21 e.u. The free energy change of hydrogen ion binding under the circumstances investigated is primarily determined by the decrease in entropy and is therefore positive. If the polymerizing unit is a double spiral, 34 subunits will bind 34 r hydrogen ions for each one which polymerizes. The increase in free energy at 290°K from hydrogen ion binding will therefore be 34 $[(-123 - (-8.35)]T)r$, where r is the number of hydrogen ions bound per protein monomer. This calculates out to be +78 kcal/mole per hydrogen ion bound per subunit. At *pH*=6.25, this figure would be 11 kcal/mole less. Since DTMVP seems to bind more hydrogen ions per mole of protein monomer upon polymerization than does TMVP, this positive contribution to the net free energy should be greater for DTMVP than for TMVP. Furthermore, this contribution in both cases is reduced when the *pH* is lowered. It is apparent, therefore, from the discussion of the present issue and the previous discussion of the electrical work term, that the differences in polymerization properties of two proteins can be explained, at least qualitatively, in terms of the differences in charge and in amount of hydrogen ion bound upon polymerization.

Even though the amino acid sequence of DTMVP differs from that of TMVP in 28 of a total of 158 amino acids, inspection of these replacements shows no net change in the number of hydrophobic groups. Furthermore, SARKAR (11) showed that the structural and bonding properties of TMVP and DTMVP must be quite similar despite the significant difference in amino acid sequence, because these two proteins will co-polymerize *in vitro* to form rods with electrophoretic mobility intermediate between those of TMVP and DTMVP. Thus, it seems reasonable to assume that the entropic contributions to polymerization are the same for the two proteins.

The situation with respect to the difference between TMVP and PM2P is not amenable to interpretation in these terms. The most remarkable difference between the polymerization in 0.1 μ phosphate buffers of PM2P and TMVP is the fact that steep $(OD - OD_0)$ vs. T curve first appears at pH=5.53, instead of pH=6.5 or higher. In the case of DTMVP, this characteristic type of polymerization first appeared at pH=6.25, rather than 6.5. As shown above, the shift in the case of DTMVP correlates with the slightly higher net negative charge on DTMVP and with the greater extent of H^+ ion binding. If the same considerations applied in the case of PM2P, a very much higher net negative charge would be required than for TMVP. However, the amino acid composition shows that PM2P has exactly the same number of positive and negative groups as TMVP, the only difference in ionizing residues being that the glutamic acid in position 95 on TMVP is replaced by aspartic acid in PM2P. Other things being equal, one would expect the same charge on the two proteins and the same amount of H^+ ion binding. If the two opposing tendencies are the same for PM2P and TMVP, then the entropic interaction should be significantly weaker for PM2P than for TMVP, so weak it cannot overcome the electrostatic repulsion and permit H^+ ion binding until the pH is reduced to the neighborhood of 5.5. However, the other amino acid residue difference between the two proteins is that threonine in TMVP is replaced by isoleucine in PM2P. But isoleucine is hydrophobic, while threonine is hydrophilic. Thus, other things being equal, interaction should be stronger in PM2P than in TMVP; this difference is in the wrong direction.

A remaining possibility is that the two amino acid replacements are sufficient to cause a substantially different folding of the two peptide chains. SENGBUSCH (49) suggested this to explain the fact that PM2P would not coat its own RNA. If this is indeed the explanation, it must be noted that in the main the ultimate structures produced by the polymerization of PM2P and TMVP are very similar, if not identical. The main difference between them is in the conditions under which these structure are formed.

Summary

The polymerization properties of the coat protein of cucumber virus 4, CV4P, investigated qualitatively rather than quantitatively, are roughly the same as those of TMVP. Treatment with succinic anhydride greatly increases the net negative charge on CV4P and not only prevents the formation of higher polymers but also causes the dissociation of the protein into the basic repeat unit.

The coat protein from Dahlemense strain of TMV, DTMVP, is similar to that of TMVP in that it contains 158 amino acid residues. But it differs with respect to 28 positions along the chain. It has one less number of negatively ionizing groups but two fewer positively ionizing groups. The polymerization behavior of DTMVP is almost identical with that of TMVP at *pH*=6.25 in 0.1 μ phosphate buffer. However, the effect of change in *pH* and of change in ionic strength upon polymerization behavior is greater for DTMVP than for TMVP. Hydrogen ion titration curves indicate that DTMVP binds somewhat more hydrogen ion upon polymerization at constant *pH* than does TMVP. The most important finding with DTMVP is that the material exists essentially in the form of particles with molecular weight comparable to or higher than that of double discs at the beginning of the rod-forming polymerization process.

The coat protein of the PM2 mutant of TMV, PM2P, differs in amino acid composition from TMVP in only two residues. Isoleucine replaces threonine at position 28, and aspartic acid replaces glutamic acid at position 95. These two replacements, however, result in a protein which has vastly different polymerization properties from those of TMVP and which is defective in that it is incapable of coating its own RNA. The late stage polymerization of PM2P at *pH*=5.5 in 0.1 μ phosphate buffer is closely similar to that of TMVP at *pH*=6.5. The starting material for late stage polymerization seems to be something with molecular weight equal to, or greater than, that of double discs. In 0.1 μ ammonium acetate buffer at *pH*=5.2, an entirely different polymerization pattern, seen only rarely with TMVP, is obtained. In this case long, thin open helices consisting of single strands, double strands, and double-double strands of protein monomers are formed.

Based on the evidence summarized above and that presented in Chap. 4, it seems probable that rod-forming polymerization of the coat proteins of TMV and its strains involves polymerization of units at least as large as double discs, such as the hypothetical double spiral. The free energy of polymerization derived from light scatter or other studies is the algebraic sum of a number of terms, some positive and some negative. The current state of our knowledge can be summarized by the equation:

$$\underset{+\ \text{or}\ -}{\Delta F_{\text{net}}} = \underset{-}{\Delta F_{\substack{\text{water}\\\text{release}}}} + \underset{+}{\Delta F_{\substack{\text{Sackur-}\\\text{Tetrode}}}} + \underset{+}{\Delta W_{\text{el}}} + \underset{+}{\Delta F_{\substack{H^+\\\text{binding}}}} + \underset{+\ \text{or}\ -}{\Delta F_{\text{unknown}}}$$

Estimates of the values of the individual terms in this equation are of the right order of magnitude to yield appropriate values of ΔF_{net}. Furthermore, the differences between the polymerization behavior of DTMVP and TMVP can be accounted for by differences in the net charge and in the number of hydrogen ions bound upon polymerization. These differences change the magnitude of the third and fourth terms on the right of the equation, and their variation with *pH* and ionic strength is in a manner consistent with the difference between the variations of ΔF_{net} with *pH* and ionic strength for the two proteins. Thus far, it has not been possible to explain the remarkable difference between PM2P and TMVP in this manner. PM2P has the same number of positive and negative groups as TMVP; the two proteins should have the same charge and the same H^+ binding upon polymerization. Since PM2P does not polymerize until the *pH* is reduced to 5.5 and since the opposing tendencies from hydrogen ion binding and electrical work should be the same for PM2P and TMVP, one can reason that the entropic interaction or water release term should be significantly weaker for PM2P than for TMVP. However, PM2P has one more hydrophobic amino acid residue than TMVP, a fact which counteracts this argument. It is possible that the two amino acid replacements in PM2P are sufficient to cause a substantially different folding of its peptide chain compared to that of TMVP. However, no specific information is available to substantiate this possibility.

Chapter 6

Reconstitution of TMV

Introduction

In his early studies on the polymerization of the degradation products of TMV, SCHRAMM (1) sometimes obtained nucleoprotein rods. Later, SCHRAMM reconstituted nucleoprotein rods from protein and RNA (2). In two cases out of fifteen, this reconstituted nucleoprotein showed an increase of biological activity over the starting materials. FRAENKEL-CONRAT and WILLIAMS (3) and then LIPPINCOTT and COMMONER (4) provided unquestionable evidence that copolymerization of TMV-RNA and TMVP formed rod-like particles, similar to or identical with TMV, exhibiting appreciable biological activity. It was originally reported that this result represented reactivation of inactive material (3). However, in 1956 two types of evidence converged to identify RNA as the sole seat of infectiousness. FRAENKEL-CONRAT (5) and GIERER and SCHRAMM, (6) found that highly purified preparations of TMV-RNA are infectious, while LAUFFER *et al.* (7) independently showed that x-ray inactivation of TMV is associated with rupture of the RNA chain and concluded that it is essential that RNA, but not necessarily protein, be intact for infectiousness of the virus. Thus, the reconstitution reaction leads to the formation of infectious TMV by reassembling infectious RNA and non-infectious protein. The most favorable conditions for reconstitution are 0.1% TMVP in 0.1 M pyrophosphate buffer at pH=7.25 at 30°C. Reconstitution requires about six hours. Yields of reconstituted particles of the order of 50% or more are frequent. The reaction can take place in other buffers but usually with lower yield (8). As will be shown in a subsequent section, reconstitution of TMV is an endothermic process.

Mixed Reconstitution

Cross reconstitution experiments were carried out with protein and RNA from different strains of TMV. As was to have been expected from the earlier finding that the RNA is the infectious principle, these studies showed that the reconstituted viruses produced diseases characteristic of the strain from which the RNA was obtained, (9). However, they have serological properties characteristic of the coat protein.

HART and SMITH (10) reported that TMVP can reconstitute with a number of synthetic polynucleotides. A more detailed investigation by FRAENKEL-CONRAT and SINGER (8) showed that the reconstitution took place with poly-A and poly-I but not with poly-C, poly-U and poly-G under conditions favorable for reconstitution with TMV. It is possible that some polynucleotides which do not reconstitute under these conditions can be trapped (10) when the protein polymerizes.

HOLOUBEK (11) extended the studies on the reconstitution of TMVP with RNA from different strains of TMV. TMV like-rods can also be obtained from TMVP and RNA derived from the spherical bacteriophage, MS2 (12). MATTHEWS and HARRIS (13) found that TMVP can reconstitute with the RNA from turnip yellow mosaic virus. In both cases, rods are formed which are very similar to TMV except for length.

Reconstitution is possible between TMV-RNA and coat proteins from spherical viruses. The protein from Cowpea chlorotic mottle virus will reconstitute with TMV-RNA to form a spherical particle with infectivity, (14). The proteins from brome mosaic virus and broad bean mottle virus will also reconstitute with TMV-RNA (15).

Synthesis of TMV in the Plant

REDDI (16) has reviewed the evidence pertaining to the production of TMV in the infected plant. As was pointed out in the introduction, TMV-RNA is the infectious principle. When a healthy plant cell is infected with TMV, the RNA finds its way to the nucleus. All available evidence seems to indicate that replication of the RNA takes place in the nucleus. This conclusion is supported by cytological studies using specific stains for RNA (17, 18), by microradiographic techniques (19) and by the demonstration by autoradiographic methods of virus-specific incorporation of radioactive uridine into the nucleus (20). The mechanism of replication of the RNA seems to be similar to that for RNA animal viruses. NILSSON-TILLGREN (21) has isolated both a double-stranded replicative form of TMV-RNA and also a multi-stranded replicative intermediate form.

REDDI (16) concludes that virus protein is also synthesized in the nucleus, because viral RNA must provide the template for virus protein synthesis and viral RNA is found only in the nucleus. Furthermore, SHALLA and AMICI (22) used ferritin-congugated antibodies to TMVP to localize virus protein in TMV-infected cells of tomato leaves. With the electron microscope, they showed that some ferritin-labeled TMVP was found in nuclei, but never in the form of complete virus rods.

The final question concerns where the protein combines with the RNA to form the complete virus particle. REDDI (16) presents arguments supporting the conclusion that this, too, occurs in the nucleus. REDDI (23) made homogenates of infected tobacco leaves grown in a medium containing ^{32}P. These were then fractionated. Nucleoprotein particles which looked exactly like TMV were demonstrated in the fraction containing the nuclear material. The chloroplast fraction was also examined and found to be free of TMV particles. Cytologists, however, question the validity of results of this sort because organelles isolated by fractionation procedures are contaminated by material absorbed onto them.

By the method described above, SHALLA and AMICI showed that most of the ferritin granules were found in the ground cytoplasm attached to whole virus particles. There were also regions in the cytoplasm in which the ferritin granules were not attached to virus particles. This was interpreted to represent TMVP in an unpolymerized state. No TMVP could be detected in the chloroplasts. HIRAI and WILDMAN (23) presented radioactive phosphorus for periods of 2, 4, and 6 hours to young tobacco leaves 18 hours after inoculation with TMV. The leaves were then fractionated into a nuclear fraction, a chloroplast fraction, an enriched mitochondrial fraction and a cytoplasmic extract fraction. The virus and viral proteins were then precipitated with TMV antibody. It was found that from 85 to 93% of the precipitated radioactivity was in the cytoplasmic extract. The nuclear fraction contained only 2-5%. Similar amounts were found in the fractions containing the other organells, including the chloroplasts. They regarded it as doubtful that any significance could be attached to the small amounts of radioactive phosphorus in those fractions, because the organelle fractions were not washed before they were used for TMV extraction. That contamination could account for that found in the nucleus and chloroplast fraction is supported by the fact that no ferritin-labeled TMV was found in chloroplasts (22), even though a small amount of radioactive P is found in the chloroplast fraction. They concluded that newly synthesized TMV is found in regions devoid of organelles. More recent studies involving radioactive labeling led Singer (24) to postulate that chloroplasts are a site of virus assembly.

The disagreement with REDDI involves the question whether the small amount of virus found in the nuclear fraction is the result of contamination during the fractionation procedure or was present before fractionation. The facts obtained by electron microscopy are contradictory; while SHALLA and AMICI (22) found no virus particles in nuclei or chloroplasts, ESAU and CRONSHAW (25) present electron micrographs showing rods in both. There seems to be a general agreement that whether TMV particles are assembled in the nucleus and are immediately excreted, or whether excreted protein and RNA are assembled in the cytoplasm, TMV accumulates only in the ground cytoplasm.

Calorimetric Studies

SRINIVASAN and LAUFFER (26) found by calorimetric measurements that when one gram of TMVP at a concentration of 2.5 mg/ml and sufficient TMV-RNA to react stoichiometrically, are reconstituded in 0.1 M pyrophosphate buffer at pH=7.3 and room temperature, 0.198 cal are absorbed. Thus, the reconstitution *in vitro* of TMV is an endothermic and, therefore, an entropy-driven reaction. By making use of Hess's law, the calorimetric data can be analyzed to give the heats of reactions which cannot actually be carried out *in vitro*. Two hypothetical reactions have been singled out for such analysis. The first involves the heat of the reaction between one gram of TMVP at a concentration of 2.5 mg/ml and the stoichiometric amount of RNA to form TMV at room temperature in 0.01 μ phosphate buffer at *pH*=7.5. This hypothetical reaction should bear at least some resemblance to the actual reaction which takes place in the plant cell. The second analysis represents polymerized TMVP plus TMV-RNA to yield TMV, all in 0.01 μ phosphate buffer at *pH*=5 and at room temperature. This is a hypothetical reaction which permits one to determine the enthalpy of the bonding between the protein and RNA.

From the results shown in Fig. 10 of Chap. 4, the heat for changing one gram of polymerized protein at *pH*=5.0 in 0.01 μ phosphate buffer into one gram of unpolymerized TMVP in 0.01 μ phosphate buffer at *pH*=7.5 is -0.880 cal, when the protein concentration is 2.5 mg/ml. A separate experiment showed that no heat is absorbed when TMV-RNA is transferred from a 0.01 μ phosphate buffer at *pH*=5 to a 0.01 μ phosphate buffer at *pH*=7.5. Since the optimum conditions for reconstitution are obtained in 0.1 M pyrophosphate buffer at *pH*=7.3, the next step was to measure the thermal effects of transferring both the TMVP and the TMV-RNA from 0.01 μ phosphate buffer at *pH*=7.5 to 0.1 M pyrophosphate buffer at *pH*=7.3.

When TMVP at a concentration of 2.5 mg/ml is thus transferred to pyrophosphate buffer, 0.785 cal/g are absorbed. This result shows that there is substantial polymerization of the protein. The results of STAUFFER, *et al.* (27) showed that when TMVP in 0.1 μ phosphate buffer at *pH*=7.5 at a concentration of 50 mg/ml is simultaneously diluted to a concentration of 2 mg/ml and transferred to 0.01 μ phosphate buffer at *pH*=7.5, 0.343 cal are evolved per gram. An ultracentrifuge experiment of TMVP in 0.1 μ phosphate buffer at *pH*=7.5 at a protein concentration near 50 mg/ml showed that almost all of the material sediments in a single boundary with an uncorrected sedimentation coefficient at 23.2^{o} of 17.6 S. Since it was obviously impossible to extrapolate in the usual manner to zero concentration, it was necessary to make use of the observation of LAUFFER (28) that an equivalent result can be obtained when one corrects for solution viscosity. When this and a correction for the density of the solution were made, a value of s_w^{20} of 22.8 S was obtained. This agrees well with the normally accepted value for double discs. Data obtained in the osmometer showed that the number average molecular weight at 2 mg/ml in 0.01 μ phosphate buffer is 87,000, and at 3 mg/ml is 89,700. One can conclude that when double discs are formed by beginning with TMVP in 0.1 μ phosphate buffer at a concentration of 2 mg/ml, 0.343 cal are absorbed per g. STAUFFER *et al.* (27) also obtained data which indicate that if one begins with TMVP in 0.01 μ phosphate buffer at *pH*=7.5 at a concentration of 3 mg/ml, 0.307 cal are absorbed per g when double discs are formed. A value intermediate between these two, near 0.325 cal per g, should represent the result expected if the starting concentration is 2.5 mg/ml. Thus, the calorimetrically determined heat of transferring TMVP at a concentration of 2.5 mg/ml from 0.01 μ phosphate buffer at *pH*=7.5 to 0.1 M pyrophosphate buffer at *pH*=7.3, +0.785 cal/g, is more than twice that for formation of double discs.

When the stoichiometric amount of RNA which will combine with one gram of TMVP is transferred from 0.01 μ phosphate buffer at *pH*=7.5 to 0.1 M pyrophosphate buffer at *pH*=7.3, heat is evolved to the extent of 0.258 cal. This negative heat has been shown by SRINIVASAN and LAUFFER (29) to be correlated with a blue shift in the ORD spectrum. CANTOR, *et al.* (30), provided evidence that base-pairing leads to a similar blue shift. It is possible therefore that base-pairing is the major source of the heat evolution obtained when RNA is transferred to pyrophosphate buffer. Thus, both the protein and the RNA are in structured states in 0.1 M pyrophosphate buffer at *pH*=7.3 before they react to form TMV.

The results of SRINIVASAN and LAUFFER (26) show further that when one gram of TMVP at a concentration of 2.5 mg/ml and sufficient TMV-RNA to react stoichiometrically, both in 0.1 M pyrophosphate buffer at *pH*=7.3, are mixed at room temperature, 0.198 cal are absorbed. A separate experiment showed that transferring the resultant TMV from 0.1 M pyrophosphate buffer at *pH*=7.3 to 0.01 μ phosphate buffer at *pH*=7.5 results in no heat absorption. Finally, Fig. 10 of Chap. 4 shows that transferring one gram of TMV from 0.01 μ phosphate buffer at *pH*=7.5 to 0.01 μ phosphate buffer at *pH*=5 involves an evolution of 0.330 cal.

When the positive and negative heats for transferring one gram of TMVP at a concentration of 2.5 mg/ml and the stoichiometric amount of RNA needed to combine with it from 0.01 μ phosphate buffer at *pH*=7.5 to 0.1 M pyrophosphate buffer at *pH*=7.3, for combining the TMVP and the RNA in 0.1 M pyrophosphate at *pH*=7.3 and for transferring TMV from 0.1 M pyrosphosphate buffer at *pH*=7.3 back to 0.01 μ phosphate buffer at *pH*=7.5 are added, a value of +0.725 cal/g results. Thus, the hypothetical reaction between TMVP and its RNA in 0.01 μ phosphate buffer at *pH*=7.5 at room temperature, a reaction somewhat comparable to that which must take place in an infected plant cell, is endothermic to the extent of 0.725 cal/g of protein. Since an endothermic reaction is of necessity entropy-driven, this result justifies the inclusion of a chapter on reconstitution of TMV in this book.

When all of the above mentioned negative and positive heats are added, one obtains the final result that, for the hypothetical reaction between polymerized TMVP and TMV-RNA, both in 0.01 μ phosphate buffer at *pH*=5.0 at room temperature, 0.485 cal are evolved per g of protein. This corresponds to -8500 cal/mole of TMVP chemical repeat unit. Since there are three phosphates or three bases in TMV-RNA associated with each chemical repeat unit in the virus particle, this heat corresponds to -2800 cal/mole of phosphate group or of base in the RNA.

The enthalpy described above for the interaction between RNA and TMVP of approximately -2800 cal/mole of RNA phosphate or base is more than sufficient to account for the enhanced stability of TMV over that of the polymerized protein. When protein polymerizes in 0.1 μ phosphate buffer at *pH*=6.0, ΔF^{o} at 290°K can be calculated, as was reported in Chap. 5, to be -23 kcal/mole. If the polymerizing unit is the double disc, it would be associated with 102 RNA phosphates or bases in the TMV particle, the enthalpy of which would contribute 102x (-2.8) or -286 kcal/mole. Even if compensated in large part by $T\Delta S^{o}$, this will increase the stability of virus over polymerized protein enormously.

The bonding between RNA and protein in TMV is frequently assumed to be a salt linkage between positive amino and negative phosphate groups. At least superficial support for this idea is afforded by the result under discussion. In the interior of proteins, according to KAUZMANN (31), ΔH for salt linkages can be somewhat negative. If there are three salt linkages per monomer in TMV, at least one of them must be with an arginine amino group, because there are only two lysine amino groups.

The acid base titrations of SCHEELE and LAUFFER (32) show that between *pH*=7.3 and 5.0, TMVP binds 3.5 hydrogen ions per chemical subunit. Over the same *pH* range, TMV binds between 2.8 and 2.9 hydrogen ions per chemical repeat unit; if there were any titration of RNA (33), TMV should bind slightly more than TMVP in this *pH* range. Furthermore, the electrophoretic mobility, which is proportional to net charge on the surface, is identical for TMV and polymerized TMVP at *pH*=5.0 and also at *pH*=6.0 (34). It follows therefore that when RNA combines with TMVP, there can be no major change in net charge. Since RNA at *pH* values between 6 and 7 has approximately one negative charge per phosphate group, positive ions must be bound when TMVP reacts with RNA. SRINIVASAN and LAUFFER (26) carried out reconstitution of TMV in an unbuffered medium containing 0.15 M KCl adjusted to *pH*=7.3 with base. The protein concentration was 2.81 mg/ml, and the RNA concentration 0.0855 mg/ml. Isolation and analysis of the products of the reaction indicate about a 30% yield of reconstituted TMV. *pH* change as a function of time was measured. During the first hour there was significant increase in *pH*, indicative of hydrogen ion binding during the reconstitution. However, the *pH* gradually fell after that time to 7.4. A 0.15 M potassium chloride solution also exhibited a decrease in *pH* over the same time, but of lesser magnitude. The initial behavior supports the idea that hydrogen ions are bound during reconstitution. The subsequent decrease in *pH* with time was probably caused by carbon dioxide uptake. When this experiment was repeated by STEVENSON and LAUFFER (35) with elaborate precautions for excluding carbon dioxide, the *pH* increased from *pH*=6.5 to 7.8 on reconstitution, and remained steady as illustrated in Fig. 1. No *pH* change was observed in a control experiment with RNA alone. A control experiment with TMVP alone showed an increase in *pH* from about 6.3 to 7, which dropped to *pH*=6.5 when the temperature was lowered from 25° to 5°C. This latter result was to be expected because it is known that protein alone binds hydrogen ions when it polymerizes in the neighborhood of *pH*=6.5. Overall, this result does seem to indicate that in reconstitution some hydrogen ions are bound in excess of those bound by the protein component.

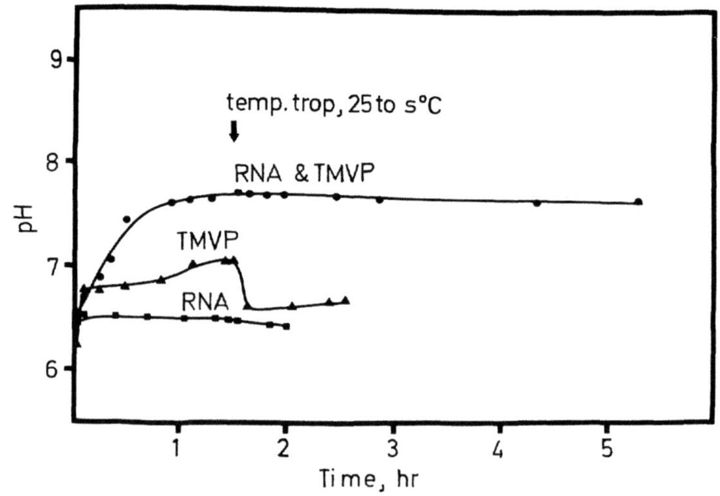

Fig. 1. Variation of *pH* with time during the reconstitution reaction. From STEVENSON and LAUFFER (35)

It has generally been assumed that reconstituted TMV is indistinguishable from TMV isolated from infected plants. ADIARTE and LAUFFER (36) discovered an important difference, illustrated in Fig. 2. When osmotic pressure measurements are made on protein solutions at *pH* values far removed from the isoelectric point and in buffers of low ionic strength, such as 0.01, one normally obtains a positive slope if osmotic pressure divided by concentration is plotted against concentration of protein. The main reason for this is the charge on the protein; this is called the Donnan effect. The Donnan effect is proportional to the slope of the graph such as that shown in Fig. 2. Since it depends upon the square of the ratio of charge to mass, for a protein of a given ratio of charge to mass the slope should be independent of molecular weight. However, the intercept of the graph should be near zero for particles of very high molecular weight like TMV. TMV isolated from infected plants shows no Donnan effect even though the ratio of charge calculated from the hydrogen ion titration curve to mass predicts that an appreciable Donnan effect should be observed. The actual situation resembles the horizontal curve of Fig. 2. However, ADIARTE and LAUFFER (36) found that reconstituted TMV did exhibit a Donnan effect of the correct order of magnitude, as indicated by the sloping line of Fig. 2. When calcium ions were added, the Donnan effect

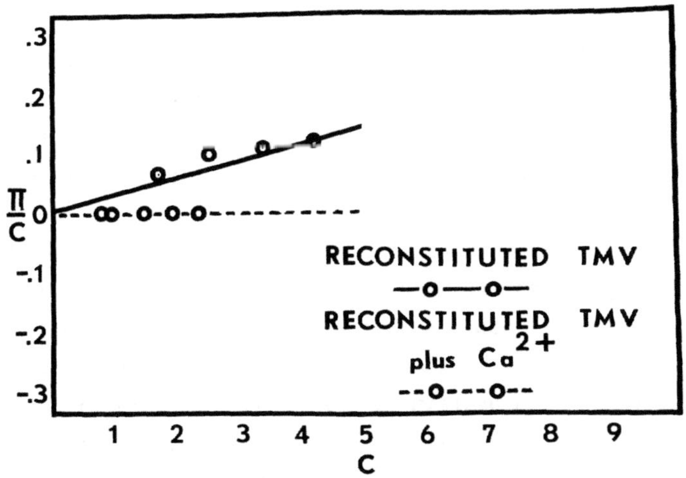

Fig. 2. Osmotic pressure plot for reconstituted TMV and reconstituted TMV treated with Ca^{2+} in phosphate buffer, ionic strength 0.01, 20.5°C. From ADIARTE and LAUFFER (36). Reprinted by permission of the copyright owner, Academic Press, Inc.

was abolished, yielding a result identical with TMV isolated from infected plants, illustrated by the horizontal line in the Figure. The explanation is that the negatively charged centers on reconstituted TMV bind calcium ions to form particles with greatly reduced net negative charge. Since the Donnan effect depends on the square of the ratio of charge to mass, it is not necessary that the charge be reduced to zero in order to render the Donnan effect immeasurably small. Since normal TMV isolated from infected plants has a Donnan effect similar to that of reconstituted TMV after binding calcium ions, it was concluded that bound calcium is a normal constituent of natural TMV, even though it is not of freshly reconstituted TMV. Thus, binding of divalent cations by virus, as well as possible binding of hydrogen ions during assembly of protein and RNA, help to explain the fact that virus and polymerized protein have indistinguishable electrophoretic mobilities. Further, ADIARTE and LAUFFER (36) also found that TMV RNA exhibits a much lower Donnan effect than expected, indicating that the negative charge on the RNA is much lower than ordinarily supposed.

Models for Reconstitution

BUTLER and KLUG (37) and BUTLER (38) have advanced the hypothesis that the double disc is the TMVP unit immediately involved in the reaction with RNA to produce reconstituted virus. The experimental basis for this hypothesis is the finding, illustrated in Fig. 3, that when one starts with protein already in the double disc form, reconstitution to form complete virus rods is completed in about five minutes, rather than in the six hours or more required when small protein aggregates are used. For these experiments, BUTLER and KLUG started with TMVP which has been dissolved in 0.1 μ phosphate buffer at *pH*=7.0 at 20°C for at least two days. DURHAM, FINCH and KLUG (39) had found that the sedimentation pattern of such protein exhibited one boundary with a sedimentation coefficient of 19 S, identified as the double disc, accounting for about 80% of the material and another with a sedimentation coefficient of 4.5 S, identified as the cyclical trimer, accounting for the balance. The 20 S protein was transferred to 0.1 M sodium pyrophosphate at *pH*=7.3 and at 25°C to yield a final concentration of 1.3 mg/ml. RNA was added to a concentration of 0.05 mg/ml. Reconstitution was estimated from the increase in turbidity at 310 mμ. When protein with a sedimentation coefficient of 4 S was added to the pyrophosphate, no appreciable increase in turbidity was observed in a period of a few minutes, but when the 20 S material was added, the increase in turbidity began immediately and achieved its maximum value in about five minutes. BUTLER and KLUG (37) state that the reason for the much slower reconstitution process observed by others is that the rate limiting step is the formation of double discs from 4 S protein. DURHAM and KLUG (40) reported that conversion of 4 S to 20 S protein requires a matter of hours or days.

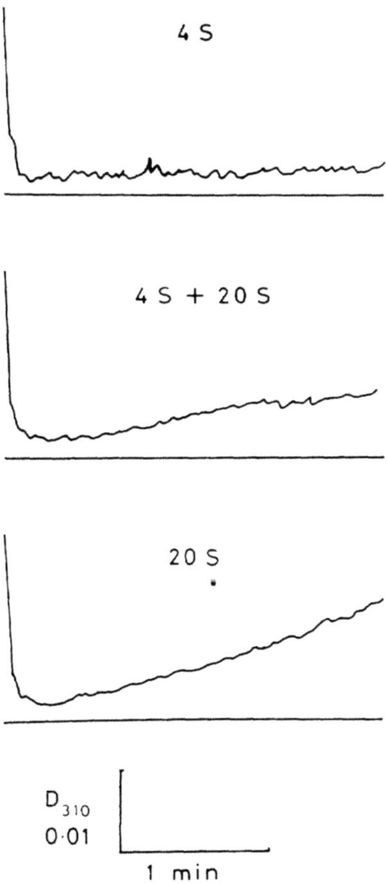

Fig. 3. Reconstitution of TMV-RNA with different protein aggregates, followed by increase in turbidity at 310 nm. Reaction in 0.1 M sodium pyrophosphate (*pH*=7.3) at 25°C. From BUTLER and KLUG (37). Reprinted by permission of the copyright owner, MacMillan (journals) Ltd.

STUSSI, *et al.* (41) found many incompletely coated RNA threads as well as completely reconstituted TMV particles in reconstitution experiments. These incompletely coated particles always contained protein on one end of the RNA thread, suggesting that TMV reconstitution is a polar process. Similar incompletely coated particles were observed by RODIONOVA *et al.* (42, 43). At pH=6.5, these particles could not be elongated into fully formed TMV, even upon prolonged incubation with protein. BUTLER and KLUG (37) then proposed further that the reaction between a double disc and an RNA chain begins from the 5'-hydroxyl end of the RNA chain. They digested TMV-RNA with calf spleen phosphodiesterase, an enzyme which attacks the 5'-hydroxyl end. They found an immediate reduction in the rate of reconstitution, followed by further reduction over a period of 80 minutes. When the RNA was digested from the 3'-hydroxyl end with snake venom phosphodiesterase, there was no appreciable change in the rate of reconstitution. Enzymatic degradation experiments reported by GUILLEY *et al.* suggested the same conclusion (44). BUTLER and KLUG (37) proposed that the first 51 nucleotides at the 5'-hydroxyl end of the RNA have an exceptionally strong affinity for the protein subunits in the first double disc. This presumably causes the first disc to change into a double spiral. When another double disc is added to the end of the resulting structure containing protein in the helical form and an RNA chain protruding from the end, it is postulated that the protein units in sequence snap into the helical form, thereby enclosing more and more of the RNA.

Further evidence that TMV-RNA adds protein from the 5'-hydroxyl end during reconstitution was presented by THOUVENEL *et al.* (45) and was reviewed by STUSSI *et al.* (46). Partially reconstituted TMV consists of two fractions, fully coated virus and virus coated with protein only about one fourth of the way from one end. They were able to separate these two components and adduce evidence that the end of the RNA thread protruding beyond the protein coat is the 3'-hydroxyl end. To do this, they made use of the fact that the 3'-hydroxyl end is not phosphorylated and thus possesses two adjacent free hydroxyl groups in the cis position. This permits oxidation by sodium periodate to open the terminal ribose ring and form a dialdehyde, which, in turn, will bind radioactive Dimedone. This is the only position on the TMV-RNA chain which can react in this way. RNA extracted from fully coated rods exhibited this reaction. However, RNA extracted from partially coated rods following digestion with pancreatic ribonuclease, which removes the exposed part of the nucleic acid thread and which leaves a fully phosphorylated end, does not bind radioactive Dimedone. When reconstitution was carried out with RNA which had

Dimedone bound to its 3'-hydroxyl end and when partially coated rods were later digested with ribonuclease, the Dimedone was contained in the release RNA fragments. Thus, these experiments show that the 3'-hydroxyl end was the end protruding and that the 5'-hydroxyl end must have been the one coated with protein.

OHNO *et al.* (47) proved the same point by a different method. First they partially stripped TMV with 1% sodium dodecyl sulfate, then digested the partially stripped virus particles with pancreatic ribonuclease. MAY and KNIGHT (48) had shown previously that such stripping begins at the 3'-hydroxyl end. Thus, the residual protein-coated RNA contained the 5' end. RNA was then isolated from these particles and was shown to reconstitute as well as full length RNA. In another experiment, they labeled TMV-RNA at the 3'-hydroxyl end with tritium. They then reconstituted and digested with pancreatic RNAase. The tritium label was removed. In other experiments they cleaved the tritiaded RNA ultrasonically and showed that the bulk of the radioactivity remained in the RNA fraction following reconstitution.

RICHARDS and WILLIAMS (49) questioned the conclusion of BUTLER and KLUG (37) that the growth of rods, once initiated, resulted from the addition of double discs, but they concurred that the initiation of rods involves the addition of RNA to discs. Similar conclusions were reached by OKADA and OHNO (50). RICHARDS and WILLIAMS reported a number of experiments, one of which is here described in detail. First, in good agreement with the previous findings (39), they observed that a solution of TMVP at a concentration of 5 mg/ml in 0.1 μ sodium phosphate buffer at *pH*=7, which had been allowed to stand at 20°C for 10-15 hours, showed that in the ultracentrifuge 60-70% of the material sedimented at 20 S and the remainder sedimented at 4 S. When such material was chilled to 0°C for periods of 15 minutes or longer, centrifugal analysis at 4-8°C showed a single 4 S boundary. On rewarming chilled material to 20°C, the original equilibrium between 20 S and 4 S was established only after many hours (40). The experiment consisted of adding 0.01 ml of TMV-RNA at a concentration of 5 mg/ml, both in 0.1 μ sodium phosphate buffer, to 0.8 ml of 0.1 M sodium pyrophosphate buffer at *pH*=7.3 and 26°C. Samples for electron microscopy were taken from the reaction mixtures at 2-3 minute intervals and diluted 50-fold in an ice-cold NH_4HCO_3 solution at *pH*=7.8 containing polystyrine latex spheres at a known concentration. Such dilution has no effect on TMV, but it depolymerized double discs and larger aggregates of TMVP to material too small to be detected in the electron microscope. Thus, electron microscopically visible material represents only TMVP bound to RNA. Droplets weré sprayed upon electron microscope grids and were shadowed with uranium. The number of visible rods per unit volume of reaction mixture was counted, and the length of each rod was measured on the electron micrograph. The results are shown in Fig. 4. Curve A represents the number of rods per 10^{-15} l as a function of time when protein which had been held for a long time at 20°C (mainly double discs) was added. It can be seen that within three minutes the number of rods formed was practically at the saturation value. Curve B shows, in contrast, that the number of rods increased only very gradually with time when TMVP which had been chilled to 0°C for at least 15 minutes, and which contained only 4 S material, was added. RICHARDS and WILLIAMS interpret the difference between these two curves to mean that RNA threads reacted very rapidly with double discs until all threads were attached to double discs in the case of curve A. However, in the case of curve B, no double discs were present at time zero but formed gradually and slowly, and only after forming were they able to react with RNA threads.

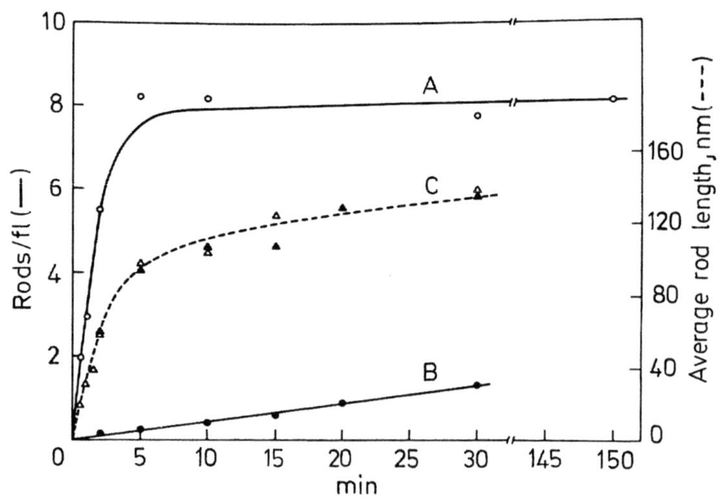

Fig. 4. TMV rod assembly: comparison of rod numbers and rod lengths. Reactions were in sodium pyrophosphate buffer. Curve A: disc protein; Curve B: A-protein; Curve C: (△) disc protein, and (▲) A-protein. From RICHARDS and WILLIAMS (49). Reprinted by permission of the authors

Curve C of Fig. 4 represents the average length of rods as a function of time. The open triangles are for the case in which the protein was added in the form of double discs plus some 4 S material, and the closed triangles represent the case in which the protein was added in the form of 4 S material. The results are indistinguishable. RICHARDS and WILLIAMS interpret this to mean that rods grow even faster when 4 S material is the predominant form than when double discs are the predominant form, because, as they point out, the rods available to grow in the former case were available for a shorter period of time on the average than in the latter case.

This experiment and others were interpreted to mean, in agreement with the conclusion of BUTLER and KLUG, that rods are initiated by the addition of RNA threads to double discs. But, in disagreement with the conclusion of BUTLER and KLUG, rods once initiated grow by the addition of either monomeric protein units or very small polymers of them. If the protein is largely in the double disc form, it first dissociates to the 4 S state before adding to rods. In rebuttal, BUTLER and KLUG (51) report that they have performed new experiments which confirm their previous results and show that the contrary findings of RICHARDS and WILLIAMS result from a non-equilibrium form of "A-protein", produced by dissociation of double discs upon cooling, which has a "memory" for the double disc state and rapidly reaggregates, either in solution or on the growing particle.

When RICHARDS and WILLIAMS (52) investigated further the elongation of partially assembled rods, they found by turbidimetric measurements and electron microscopy that the elongation rates are about the same regardless of whether protein in the disc state or in the A state was used. Furthermore, they performed their experiments with A-protein prepared by three different methods, "equilibrium" A-protein kept exclusively in this form for at least 20 hours before use, "freshly made" A-protein like that used in their earlier study prepared by cooling disc protein in 0.1 μ phosphate buffer at *pH*=7 to $0^{o}C$ for 15 minutes and "helix" A-protein prepared by depolymerizing a preparation which had been stored as TMV protein helix. All three were indistinguishable in their ability to bring about the elongation.

The fact that thermodynamic methods have been as useful as they have been in studying disc-forming polymerization of TMVP implies that there is always a dynamic equilibrium between large aggregates, even double discs, and all smaller species down to monomeric protein subunits. As will be seen in Chap. 8, there is positive evidence that the equilibrium between G-actin and F-actin is indeed dynamic. One thing that is known for sure is that the free energy of the "bond" between TMV and one of its protein subunits is vastly greater than that between a protein subunit and any reversible polymer of protein subunits. The evidence for this is that these protein polymers are ealily dissociated by slight changes in *pH* or temperature or concentration. In contrast, the virus itself can be disassembled only by drastic treatment. Therefore, if partially formed TMV rods and double discs are brought together under conditions in which the reaction between protein and nucleoprotein is kinetically possible, nucleoprotein rods will grow at the expense of double discs. This is what happens. The methods used to observe this change are those which detect either the disappearance of the double discs or the growth of the nucleoprotein rods. These methods simply do not differentiate between the binding of a small fragment by a growing TMV rod, followed by an adjustment of the equilibrium between double discs and small fragments, and the binding of a double disc, followed by necessary internal rearrangements to achieve the final structures found in the virus.

Looked at another way, if a double disc comes very close to a growing nucleoprotein rod, and if that double disc dissociates a protein monomer or a small aggregate of protein monomers, as it must do if it is in dynamic equlibrium, it is much more likely that that fragment will be captured by the growing nucleoprotein rod than that it will return to the double disc. The same can be said for subsequent fragments that dissociate off. Thus, it would be possible that all of the protein subunits from a double disc could add to a particular growing rod in a very short time. It is now a matter of semantics whether one describes this as the addition of a double disc or the addition of smaller fragments.

Evidence obtained by OKADA *et al.* (53) and OHNO *et al.* (54) supports the idea that once partially assembled rods are formed, they grow by adding particles smaller than double discs. When TMVP is oxidized by N-bromosuccinimide, tyrosine residue 139 is oxidized to dibromo tyrosine (54). This altered protein, NBS-TMVP, can be polymerized at *pH*=6.5 into long rods resembling intact virus when the temperature is raised from 5 to 30°C. In 0.1 M pyrophosphate buffer at *pH*=7.2, it will form an 8 S component but no double discs. Neither will it undergo reconstitution (53). However, NBS-TMVP can add to partially assembled rods which were formed from the reaction of a small amount of normal protein in the double disc state with RNA. By this process, fully coated active virus particles are formed (54). In this case, the growth could not have been the result of adding double discs; none were present.

Cucumber green mottle mosaic virus protein, CGMMVP, described in Chap. 4, can be reconstituted with TMV RNA to form a hybrid virus (OHNO *et al.* 55). However, this protein forms only single discs in 0.1 M pyrophosphate buffer at *pH*=7.2. These single discs are required to initiate reconstitution, that is, to form partially assembled rods. However, reconstitution does not proceed to completion in the presence of these discs. Only after the buffer is diluted to 0.05 M, where the protein is in the 2.5 S state, probably corresponding to monomers in equilibrium with dimers, do the partially assembled rods grow to form complete hybrid virus particles. Here again, small units exclusively are involved in the elongation process.

A nucleotide sequence of part of the strand of TMV-RNA has been reported (56). TMV-RNA was partially digested with ribonuclease T_1 for 30 minutes at $0^{\circ}C$. The fragmented RNA, after phenol extraction to remove the enzyme, was reconstituted with TMV protein in 0.5 ionic strength, *pH*=7.25 pyrophosphate buffer. Reconstitution was allowed to proceed for four hours at $24^{\circ}C$ with a protein-RNA ratio of 1:2. The nucleoprotein produced was purified by centrifugation. It appeared in the electron microscope as discs and very short rods. The RNA was extracted from this nucleoprotein component and sequenced. The results as reported (56) is:

AAGUUGAAAAUCAGGCGAAC (C,C,AC) GACUGCCGAAA
CGUUAGAUGCUAC (U,C) GUAGAGUAGACGACGACGCGG
UGGCCAUAAGGAGCGAUAAAU (AAU,UUU) AAUAG

Reconstitution of Other Viruses

MORRIS and SEMANCEK, (57) described the reconstitution of the C strain of tobacco rattle virus. As was pointed out in Chap. 5, this is a complex virus consisting of at least two types of rod shaped particle, one long and one short. The protein is common to both; the difference lies in two types of RNA. Reconstitution takes place best at low temperatures, between 4 and $10^{\circ}C$, at high *pH* values between 7.5 and 8.0, at ionic strength around 0.01 and in media of high dielectric constant achieved by adding lysine to a concentration of 0.25 m. These conditions for optimal reconstitution are roughly those under which double discs with a sedimentation coefficient of 35 S are found. Thus, double discs seem to be essential for at least some stage of the reconstitution process.

HAIDAR *et al.* (58) studied the reconstitution of the CAM strain of the same virus. Unlike the C strain, this virus reconstitutes best at *pH*=4.7 in 0.5 m phosphate buffer at $1^{\circ}C$. Electron micrographs show double discs attached to one end of partially assembled rods. No appreciable reconstitution took place at *pH*=8 in 0.25 M glycine. The CAM strain is known to differ appreciably from the C strain in amino acid composition.

Barley stripe mosaic virus, another rod-shaped virus unrelated to TMV can also be reconstituted. ATABEKOV *et al.* (59) reported that it assembled in 0.01 M Tris-HCl buffer at *pH*=7.2 - 7.5.

Summary

Biologically active TMV can be reconstituted *in vitro* from purified TMV-RNA and purified coat protein. The reaction proceeds best in a medium consisting of 0.1 M pyrophosphate buffer at *pH*=7.25 and at about $30^{\circ}C$. Cross reconstitution can be achieved with protein and RNA from different strains of TMV. The reconstituted viruses produce diseases characteristic of the strain from which the RNA was obtained, but they have serological properties characteristic of the strain from which the coat protein was obtained. TMVP will reconstitute with a number of different kinds of RNA, and TMV-RNA can reconstitute with the coat proteins from several spherical viruses.

In the infected cell, replication of TMV-RNA takes place in the nucleus. The coat protein is also synthesized in the nucleus. The question of where in the plant cells the coat protein and RNA combine to form virus is controversial. A small amount of virus is found in nuclear fractions isolated from diseased cells, but there is disagreement as to whether this is the result of contamination or of synthesis within the nucleus. There is general agreement that whether TMV particles are assembled in the nucleus and immediately excreted or are assembled in the cytoplasm from excreted protein and RNA, accumulation of the virus occurs only in the cytoplasm.

Calorimetric studies show that *in vitro* reconstitution of TMV is an endothermic and, therefore, an entropy-driven reaction. For the actual reaction at room temperature between one gram of protein and the stoichiometric amount of RNA in 0.1 M pyrophosphate buffer at *pH*=7.3, the heat absorbed is 0.198 cal. Through the use of Hess's law and auxiliary calorimetric data, one can derive the result that 0.725 cal are absorbed when a gram of protein reacts with the stoichiometric amount of RNA in 0.01 μ phosphate buffer at *pH*=7.5, a hypothetical *in vitro* reaction somewhat resembling what must take place when virus is assembled *in vivo*. Similarly, for the hypothetical reaction between polymerized TMVP and TMV-RNA in 0.01 μ phosphate buffer at *pH*=5.0 and at room temperature, 0.485 cal are evolved per gram of protein. Thus, the actual interaction between RNA and protein is an exothermic step in an overall endothermic reaction. The value obtained corresponds to -2.8 kcal/mole of phosphate or of base in the RNA. This negative enthalpy for the interaction of RNA with protein is large enough to contribute enormously to the physical stability of TMV.

Reconstitution of TMV *in vitro* seems to involve two steps. The first is the reaction of one double disc of protein with an RNA chain, which takes place at the 5'-hydroxyl end of the chain. The second step is the growth of such short rods or nuclei by the further addition of protein. BUTLER and KLUG proposed that the growth takes place by the addition of double discs. However, RICHARDS and WILLIAMS and also OKADA and OHNO provided evidence which seems to indicate that protein monomers or small polymers are involved in the growth step. This implies that, if the protein is actually in the state of double discs, it dissociates prior to contributing to the growth of virus rods. Overall, evidence from a number of laboratories seems to indicate that the elongation step can take place with protein confronting the partially formed rods in either the double disc state or some highly dissociated state.

In common with TMV, tobacco rattle virus and barley stripe mosaic virus can also be reconstituted.

Chapter 7

Sickle Cell Anemia Hemoglobin

Introduction

Sickle cell anemia is a hereditary disease transmitted in accordance with simple Mendelian laws, prevalent in certain African populations and their descendants. Individuals homozygous for the recessive gene suffer anemia, often fatal at an early age. While such individuals rarely have off-spring, the recessive gene is protected, occurring sometimes in 40% of a population. The reason is that individuals heterozygous with respect to this gene have an adequate oxygen-carrying faculty and, in addition, have resistance to malaria. Thus, in parts of the world where malaria is an ever-present health hazard, persons heterozygous with respect to the sickle cell gene have a reproductive advantage over individuals homozygous for the dominant allele.

The name of the disease derives from the fact that in deoxygenated blood of sufferers, the erythrocytes assume the shape of crescents or sickles or holly leaves. In oxygenated blood the cells have the normal shape of biconcave discs. PAULING (1) proposed in 1955 that the "sickling" behavior could be the result of a type of hemoglobin molecule chemically different from the normal one, which, upon deoxygenation, aggregates into rods, thereby twisting the red-cell out of shape. Today it is known that this interpretation is correct. The normal hemoglobin molecule consists of four polypeptide chains, each associated with a heme group. There are two kinds of polypeptide, the alpha chain, containing 141 amino acid residues and the beta chain containing 146 residues (2, 3, 4). In sickle cell hemoglobin, the two alpha chains are identical with those in normal hemoglobin, but a single amino acid replacement is found in each of the two beta chains. The sixth residue from the amino-terminal end is glutamic acid in normal hemoglobin and valine in sickle cell hemoglobin (5, 6).

Polymerization of Sickle Cell Hemoglobin

HARRIS (7) showed that when an aqueous solution of sickle cell hemoglobin at a concentration of 15.2 g/100 ml was changed from 100% oxygen saturation to 10% oxygen saturation, there was a five-fold increase in viscosity at 37.5°C relative to that of distilled water. No similar change in viscosity was observed as a result of oxygen depletion when the concentration of sickle cell

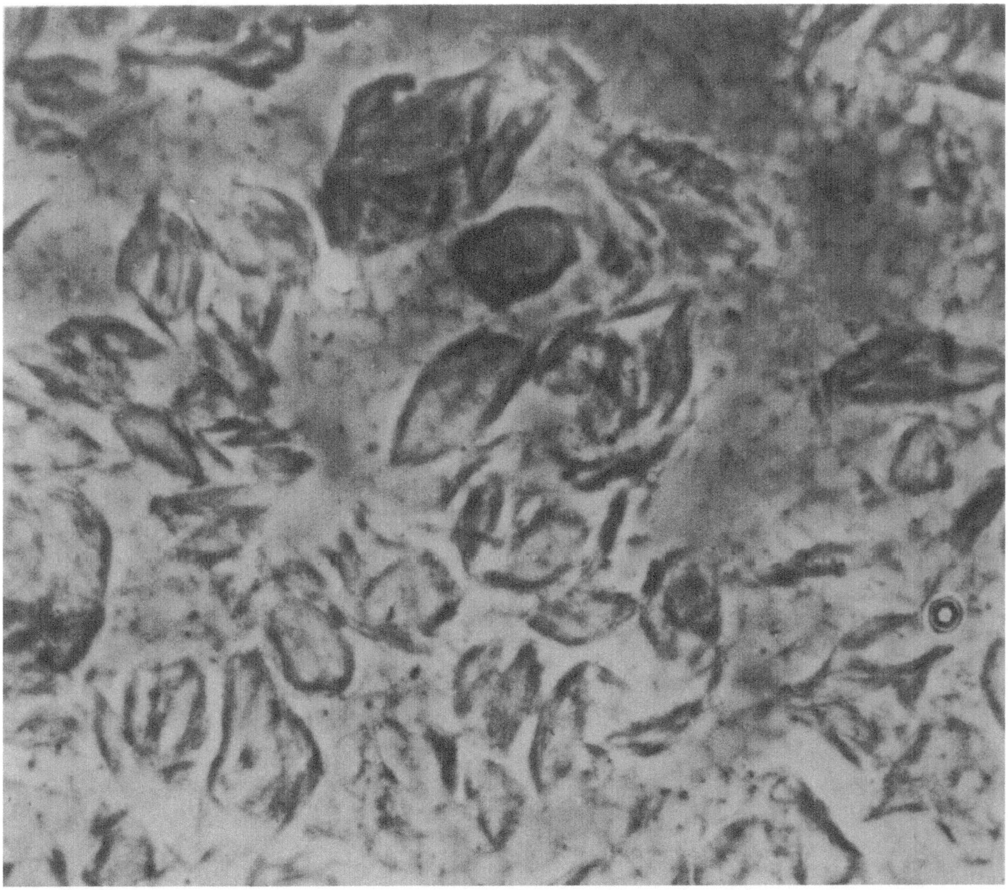

Fig. 1. Hemoglobin tactoids formed in stroma-free solutions of deoxygenated sickle cell anemia hemoglobin. (Phase microphotography X375). From HARRIS (7). Reprinted by permission of the copyright owner, the Society for Experimental Biology and Medicine

hemoglobin was decreased to 10 g/100 ml or less. Also, no comparable effect was observed with normal hemoglobin at any concentration. HARRIS interpreted this result to be evidence of polymerization of the individual sickle cell hemoglobin molecules to form elongated aggregates. While this interpretation is supported by other evidence, the viscosity experiments themselves are marred by the absence of electrolyte in the solution; abnormally high viscosities, interpretable in terms of the electroviscous effect, are common in solution of charged macromolecules free of small ions. However, HARRIS found that birefringent microtactoids, illustrated in Fig. 1, varying in length from 1 to 15 μ, were also present in the deoxygenated solutions of sickle cell hemoglobin which showed the high viscosity. Such microtactoids were absent in

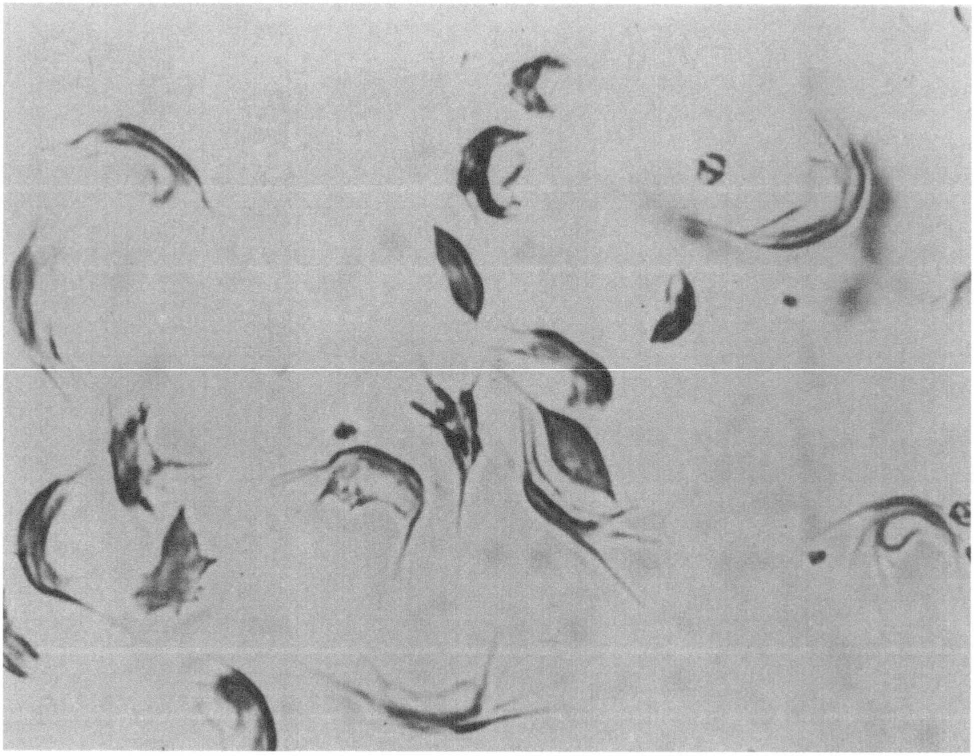

Fig. 2. Sickled erythrocytes in oxygen unsaturated whole blood from a patient with sickle cell anemia, demonstrating the similarities in shape to that of tactoids formed in stroma-free solutions of their deoxygenated hemoglobin. (Phase microphotography X375). From HARRIS (7). Reprinted by permission of the copyright owner, the Society for Experimental Biology and Medicine

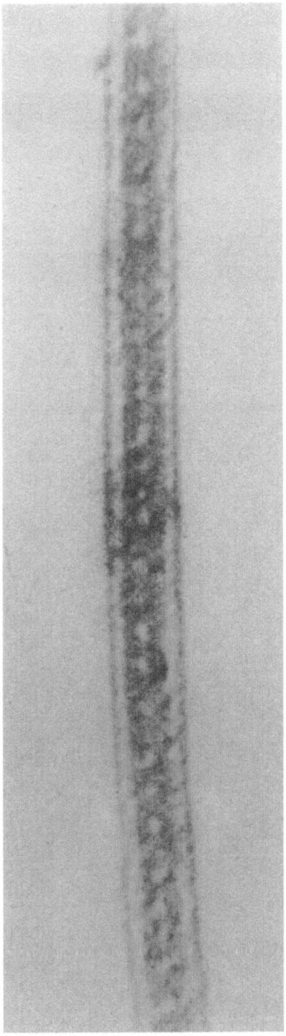

Fig. 3

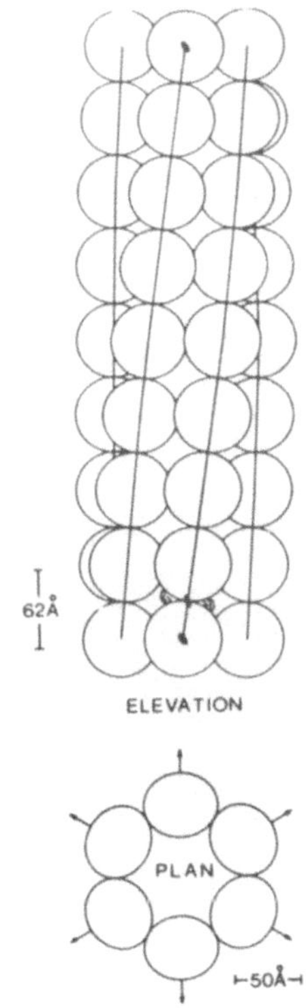

Fig. 4

Fig. 3. Electron micrograph of S hemoglobin "microtubule" interpreted as a hollow small tubule. The outer diameter of the "microtubule" is 170 Å, and the inner diameter is 40 Å. From MURAYAMA (8). Reprinted by permission of the copyright owner. Copyright 1966 by the American Association of the Advancement of Science

Fig. 4. Diagramatic representation of the structure of the hollow cable of deoxy sickle cell hemoglobin. Arrows indicate the probable positions of the molecular diads normal to the fiber axis. From FINCH *et al.* (10). Reprinted by permission of the authors

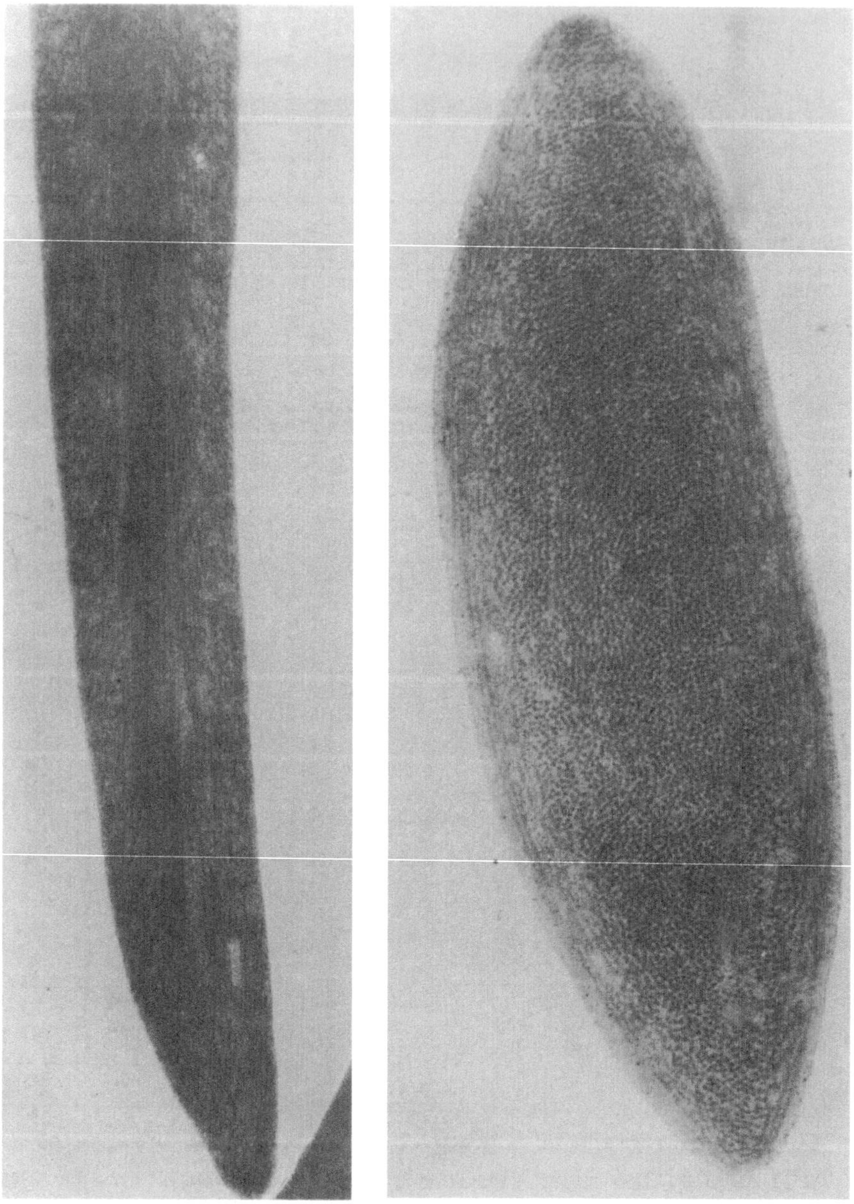

a b

Fig. 5a and b. (a) Longitudinal section of an irreversibly sickled cell x 26,800. (b) Cross-section, x 57,600. From DÖBLER and BERTLES (12). Reprinted by permission of the copyright owner, the Rockefeller University Press

normal hemoglobin, regardless of the degree of oxygen saturation, and in sickle cell anemia hemoglobin at 100% oxygen saturation. The microtactoids are of the type obtained when rod-like particles in suspension align themselves parallel to each other. Thus, the conclusion that sickle cell hemoglobin molecules aggregate into rods when oxygen saturation is reduced was valid. HARRIS observed further that at a concentration of 23.5 g/100 ml, sickle cell anemia hemoglobin formed a semi-solid gel at 10% oxygen saturation, but not at 100% oxygen saturation. The birefringent microtactoids are also present in the gels. Because these microtactoids are remarkably similar in appearance to sickled erythrocytes, which are also birefringent, Figure 2, HARRIS suggested that the sickled erythrocytes are microtactoids with membranes or membrane fragments surrounding them.

Two subsequent developments confirm and extend this interpretation. MURAYAMA (8) demonstrated double refraction of flow in solution of reduced sickle cell anemia hemoglobin and obtained pictures of sickle cell hemoglobin threads with the electron microscope (Fig. 3). They look somewhat like the microtubules found in many tissues. MURAYAMA postulated that each microtubule-like structure is composed of six hypothetical sickle cell hemoglobin monofilaments with a diameter of 65 Å, twisted together to form a hollow molecular cable with an outer diameter of approximately 170 Å and an inner diameter of approximately 40 Å. Subsequent studies by X-ray diffraction (9) and electron microscopy (10) have confirmed the essential features of this interpretation. Fig. 4 from FINCH *et al.* (10) illustrates the current understanding. STETSON (11) and DÖBLER and BERTLES (12) found rods in parallel sections and a regular array of hemoglobin in ultrathin cross sections of sickled erythrocytes, Fig. 5.

The second subsequent development, of major concern to us, is the finding by MURAYAMA (13) that, while a deoxygenated sickle cell hemolysate of sufficient concentration forms a gel at 38°C, the gel liquefies reversibly upon reducing the temperature to 0°C. Thus, gelation and, therefore, the polymerization of sickle cell hemoglobin into microtubule-like structures responsible for gelation are endothermic, necessarily entropy-driven, processes in some ways comparable to the polymerization of TMVP. MURAYAMA (14) has also reported polymerization studies on purified sickle cell hemoglobin at a concentration of 4 mg/ml in 0.2 M BIStris buffer at *pH*=6.7. Polymerization at various temperatures was followed by optical density (turbidity) measurements. Results are shown in Fig. 6. The data show that polymerization is reversible. With the exception of the lowest optical density recorded, apparently the result of experimental error, and the two values obtained at the highest temperatures, the data of Fig. 6 can be fitted reasonably well to a graph of Eq. (7) of Chap. 4. ΔH^{o} and ΔS^{o} were thus evaluated to be approximately +65 kcal/mole and +256 e.u., respectively. MURAYAMA (14) also carried out experiments on the rate of polymerization as a function of temperature. The enthalpy and entropy of activation[1] for the rate process were calculated to be +17.3 kcal/mole and +55 e.u., respectively. In common with TMVP, the polymerization is accompanied by an increase in partial specific volume. This was not measured directly for sickle cell hemoglobin, but can be inferred by the use of Eq. (15) of Chap. 1 from the observation of MURAYAMA (8) that sickled erythrocytes, which so closely resemble microtactoids, become spherical under pressures of 200 to 300 atmospheres and resume the sickle shape when decompressed slowly. However, recent studies (15) indicate that the effect of pressure is more complex than originally indicated.

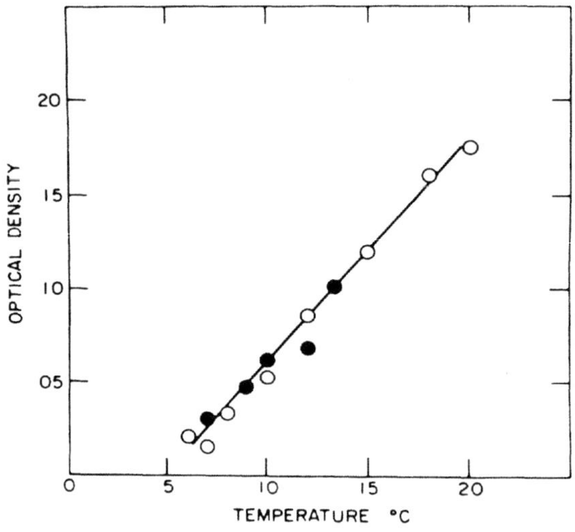

Fig. 6. Reversible aggregation of sickle cell hemoglobin in 0.05 M BIStris buffer at pH=6.7. ○ rising temperature; ● falling temperature. From MURAYAMA (14). Reprinted by permission of the copyright owner, Plenum Publishing Corp.

[1] A discussion of the concept of activation is found in Chap. 9 in the section on Protein Denaturation.

It was pointed out in Chapter 4 that when chemicals regarded to be water-structure breakers are added to the solution, TMVP polymerization is hindered or depolymerization is favored. MURAYAMA (16) dissolved propane, ethane, and methane under slight pressure in blood specimens from patients with sickle anemia. In untreated aliquot specimens, sickling was found when the blood was deoxygenated. No sickling was observed when the blood was treated with propane and almost none when it was treated with methane or ethane. These hydrocarbon molecules also inhibit gel formation. In contrast with the molecules which inhibit the polymerization of TMVP, these hydrocarbons are structure-makers in aqueous solution; at any rate, there is an increase in entropy when they are transferred from an aqueous solution to an organic solution. The probable explanation of their action is that they compete with hydrophobic centers on the sickle cell hemoglobin molecule. Oxygen, carbon monoxide (17), alkyl isocyanides, nitrosobenezene, and also sulfhydryl blocking reagents such as heavy metal ions, para-chloromercuribenzoate, N-ethylmalimide (18), iodoacetamide, iodosobenzoate, sodium maleate and ortho-chloromercuribenzoate (19), also inhibit sickling.

Sulfhydryl compounds such as H_2S, 2,3-dimercaptopropanol, and cysteine cause sickling (19). Oxygen inhibits sickling produced by any means, including use of sulfhydryl compounds (13). MURAYAMA (20), following the original suggestion by PAULING *et al* (17), interpreted the effects of these agents which prevent sickling in terms of steric hindrance.

When a sickle cell hemolysate is dialyzed against distilled water, it loses its property of forming a gel at $38^{o}C$ (21). When undialyzed deoxygenated sickle cell hemolysates are titrated with silver salts at $0^{o}C$, four silver ions are bound per four iron atoms at $0^{o}C$, indicating the availability of four free sulfhydryl groups. When the temperature is raised to 38°, only two are titrated. However, if the sickle cell hemoglobin is first dialyzed against water, then it still exhibits four free sulfhydryl groups at 0° but eight at 38°. This result was interpreted by MURAYAMA (13) as indicating a temperature mediated change in conformation of the dialyzed sickle cell hemoglobin molecule. Normal hemoglobin behaves similarily, but to a lesser extent. The specific change in conformation revealed by this study can have no direct relevance to gel formation, because dialyzed sickle cell hemoglobin does not form gels.

Additional information is available which contributes to our understanding of the organization of the hemoglobin in sickled cells, microtactoids and gels. MURAYAMA, OLSON and JENNINGS (22) concluded from dichroic-ratio measurements of sickled erythrocytes that the hemoglobin molecules are stacked along the long axis of the sickles with their heme groups essentially parallel to the long axis of the erythrocytes. Entirely consistent with this conclusions MURAYAMA (23) showed that in a magnetic field of 3.5 kG, sickled erythrocytes orient with their long axes perpendicular to the magnetic lines of force. He found that normal erythrocytes do not orient in such a magnetic field.

Optical Rotatory Dispersion

MURAYAMA (18) published optical rotatory dispersion (ORD) curves in the green region of the spectrum at 0° and $38^{o}C$ for both oxygenated and deoxygenated sickle cell hemolysates at concentrations of 11 g per 100 ml. The data are shown in Figs. 7 and 8. The patterns of oxygenated and deoxygenated material are quite different, a result probably associated with the structural shift accompanying oxygen binding. In the case of the deoxygenated hemolysate, which will form a gel upon raising the temperature from 0° to $38^{o}C$, there is a major increase in the amplitude of the positive Cotton effect when temperature is raised; this change is reduced in the presence of propane (16). The oxygenated hemolysate, which will not form a gel, exhibits a multiple Cotton effect, and the amplitude in the region of 560 mμ also increase when the temperature is raised to $38^{o}C$. Preparations of oxygenated and deoxygenated normal hemoglobin exhibit patterns essentially identical to those shown in Figs. 7 and 8 (15).

Changes in optical rotatory dispersion are conventionally interpreted in terms of changes in the conformations of the structures surrounding the light-absorbing groups. Temperature mediated changes in optical rotatory dispersion have also been observed for TMVP (24, 25, 26). However, it was difficult to decide in the case of TMVP whether any change in conformation is a prerequisite for later polymerization or is brought about as a result of polymerization. The fact that normal hemoglobin, which does not polymerize in the manner of sickle cell hemoglobin, exhibits the same temperature shift in ORD pattern as does sickle cell hemoglobin indicates that in this case the temperature

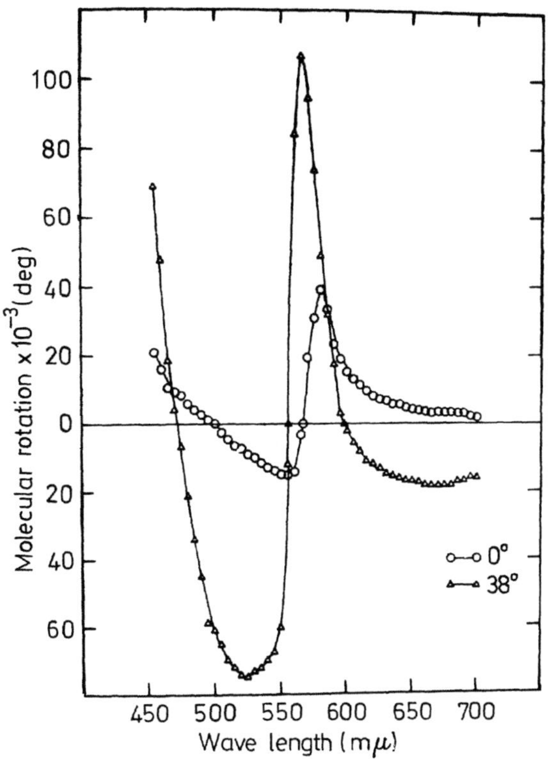

Fig. 7. Optical rotatory dispersion curves for deoxygenated sickle cell hemolysate at 0° and 38°C. From MURAYAMA (18). Reprinted by permission of the copyright owner, MacMillan (journals) Ltd.

mediated shift is certainly not the result of polymerization. The possibilities remain that it is a prerequisite or that it is independent of polymerization. The result with propane (16) slightly favors the first possiblity.

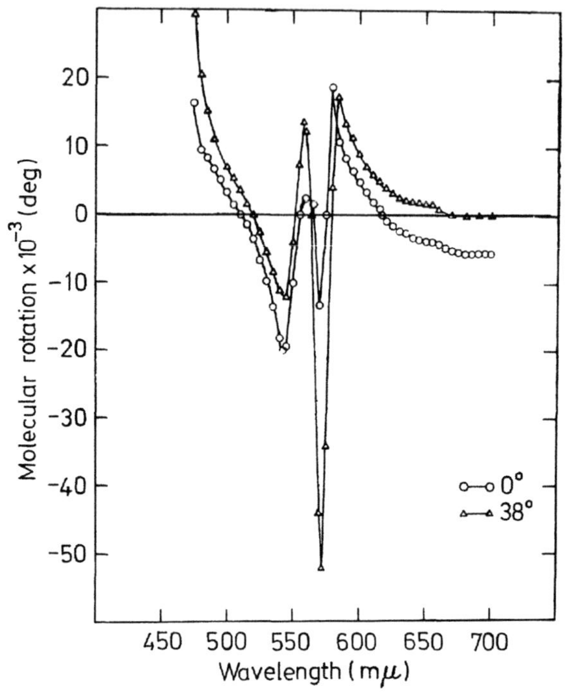

Fig. 8. Optical rotatory dispersion curves for oxygenated sickle cell hemolysate at 0° and 38°C. From MURAYAMA (18). Reprinted by permission of the copyright owner, MacMillan (journals) Ltd.

The Mechanism of Sickling

As a result of his extensive studies, MURAYAMA (8) proposed that the first step in the formation of sickled cells is a conformational change in the sickle cell hemoglobin molecule, brought about by deoxygenation, to bring it into a form which can polymerize linearly. In support of this idea, he cited the crystallographic evidence of MUIRHEAD and PERUTZ (27) that the beta chains of normal hemoglobin move together seven angstroms upon oxygenation. In a recent theoretical interpretation, he retains the idea that the reason oxygenated sickle cell hemoglobin will not polymerize, but the deoxy form will, is a matter of conformation (14).

The first eight amino acid residues from the amino-terminal end of the beta chains for normal hemoglobin and sickle cell hemoglobin are, respectively,

VAL·HIS·LEU·THR·PRO·GLU·GLU·LYS
VAL·HIS·LEU·THR·PRO·VAL·GLU·LYS

MURAYAMA's interpretation (14) is that the process of sickling takes place by entropic unions between the hydrophobic valyl residue at position 6 on the β-chain with complementary sites on the α-chain of an adjacent molecule. This stacking along the dyad axis of symmetry results in a sickle cell hemoglobin monofilament. Evidence for such a monofilament has been demonstrated on a Langmiur trough (15). MURAYAMA proposes further that six of these monofilaments twist themselves into a hollow molecular cable with outer dimensions of about 170 Å and an inner diameter of 40 Å. The existence of such cables with the right demensions has been demonstrated with the electron microscope, as shown in Figs. 3 and 5. The final step is the aggregation of such rod-like cables into microtactoids with the collapse of the red cell membrane.

FINCH *et al.* (10) ascribed a different function to the valine residue which replaces the glutamic residue at position 6 on the β-chain. They point out that it is on the outside of the hemoglobin molecule and suggest that its role is not that of changing the mode of aggregation of deoxyhemoglobin into filaments but, instead, one of stabilizing a normal contact between molecules in the deoxy-structure, possibly by promoting crystalization of the single filaments into helical cables. Whichever interpretation is correct, the role of hydrophobic valyl residues in the formation of entropic unions is clearly demonstrated.

One detail of MURAYAMA's polymerization scheme was brought into question by the studies by WHITE and HEAGAN (28). They obtained gels by mixing hemoglobin solutions at concentrations of the order of magnitude of 10% with five times their volume of 2.8 M potassium phosphate at a *pH* value near 7. Gels were obtained in this manner with normal hemoglobin and sickle cell hemoglobin, both in the oxygenated and in the deoxygenated state. Examination with the electron microscope revealed microtubules in the gels obtained from normal hemoglobin and also in the gels obtained with oxygenated sickle cell hemoglobin. The reduced sickle cell hemoglobin exhibited solid rods of diameter 240 A^{O} rather than hollow tubules. WHITE and HEAGAN therefore proposed that solid rods, not microtubule-like structures, are effective in the sickling phenomenon. FINCH *et al.* (10), however, confirmed the smaller hollow fibers. They also confirmed the occurrence of similar filaments in deoxygenated normal cells, but in much lower concentration and more irregularly aggregated. This finding is adduced to support their view that the substitution of a glutamic acid residue by valine in position 6 on the β-chain does not change the mode of aggregation of deoxyhemoglobin into filaments but, instead, stabilizes a normal contact between molecules in the deoxy structure. Sickling of red cells also occurs in the blood of the deer, the goat and the sheep (29).

Summary

Sickle cell hemoglobin differs from normal hemoglobin with respect to a single amino acid replacement in each of the two β-chains; glutamic acid residue 6 is replaced by valine. In the absence of oxygen, sickle cell hemoglobin polymerizes to form birefringent gels. Gel formation is strongly correlated with the transformation of red cells from the normal shape to the sickle shape. Like the polymerization of TMVP, sickle cell hemoglobin polymerization, gelation and sickling are endothermic processes; the reaction proceeds in a forward direction at $38^{\circ}C$, and is reversed at $0^{\circ}C$. Furthermore, the process is reversed by application of pressures in the range of 200 - 300 atmospheres. Therefore, polymerization and sickling are accompanied by an increase in volume. In addition to oxygen, many chemicals, including carbonmonoxide, simple hydrocarbons and sulfhydryl blocking agents inhibit gelation and sickling; sulfhydryl compounds promote sickling.

The mechanism of sickling involves, in the view of MURAYAMA, a conformational change in the sickle cell hemoglobin molecule to bring it into a form which can polymerize by head-to-tail stacking, involving the β-6-valine residue, along the dyad axis to form a monofilament. Six of these monofilaments then twist themselves into a hollow molecular cable with an outer dimension of 170 Å and an inner diameter of 40 Å. Finally, these cables aggregate into microtactoids and the red cell membrane collapses. An alternative view expressed by FINCH *et al.* is that the β-6-valine residue is involved in binding monofilaments together in the hollow cables.

Chapter 8

Structural and Contractile Proteins

Collagen

Collagen, the principal constituent of tendon and skin, is one of the most abundant proteins in higher vertebrates. In the animal body it is encountered primarily in the form of fibrils. These are linear polymers of a basic subunit, tropocollagen. The tropocollagen molecule consists of three polypeptide chains, each containing approximately 1,000 amino acid residues and having a molecular weight of approximately 120,000. Collagen is distinctive in that approximately 1/3 of all of its amino acid residues are glycine and approximately 1/4 are proline or hydroxyproline. Because of the presence of so much proline and hydroxyproline, the amino acid chains are not in the α-helix con-

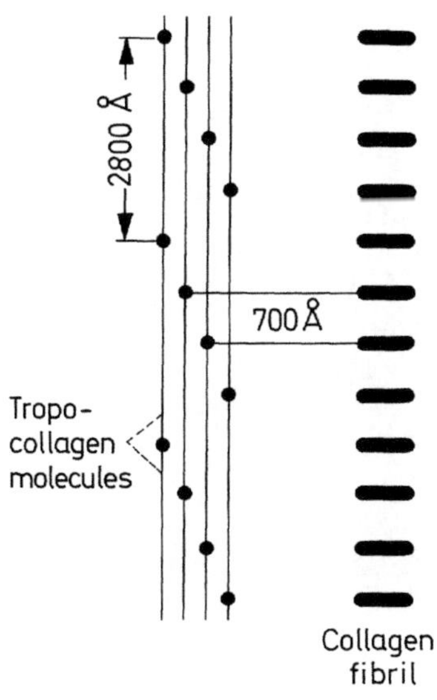

Fig. 1. Diagram showing the relationship between the 700 Å repeat units in collagen fibers and the staggered arrangement of the heads of tropocollagen molecules

figuration but in a special type of helix with a spacing of 28.6 Å. Three polypeptide chains twist together, more or less like a rope, to form a three-stranded tropocollagen molecule, 2800 Å long and 14 Å in diameter, with a molecular weight of approximately 360,000. The tropocollagen molecule can be denatured thermally; the helical structure collapses and, to a certain degree, the strands separate. At temperatures below the denaturation point, native tropocollagen molecules can aggregate to form striated collagen fibers. As illustrated in Fig. 1, tropocollagen molecules line up side by side with successive ends of adjacent molecules displaced in the fiber direction by approximately 700 Å. This produces a fibril with 700 Å spacings (1), as shown in Fig. 2.

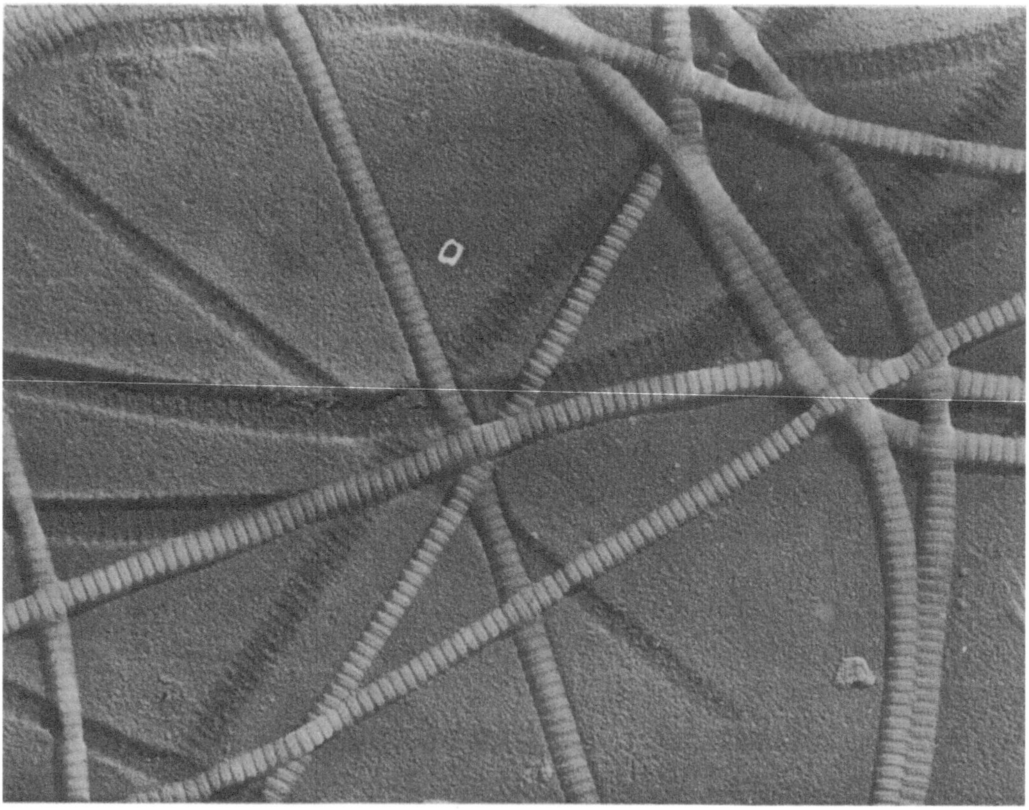

Fig. 2. Electron micrograph of collagen fibrils. From GROSS and SCHMITT (1). Reprinted by permission of the copyright owner, The Rockefeller University Press

The polymerization under various conditions of *pH* temperature, ionic strength, ATP concentration, etc. was reviewed by HARRINGTON and von HIPPEL (2). FESSLER, (3) showed that tropocollagen obtained by extracting rabbit skin with neutral salt solutions was soluble at 4°C but forms typical collagen fibers at 37°C, which disperse again upon reduction of the temperature. The effect of temperature is illustrated in Fig. 3. The relatively rigid gel obtained in this manner displays in the electron microscope a structure like that observed in fibrous collagen from tendon and hide (4). CASSEL *et al.* (5) concluded that the polymerization was a phase transition involving no change in conformation. Dilatometric studies carried out by CASSEL and CHRISTENSEN (6) showed that

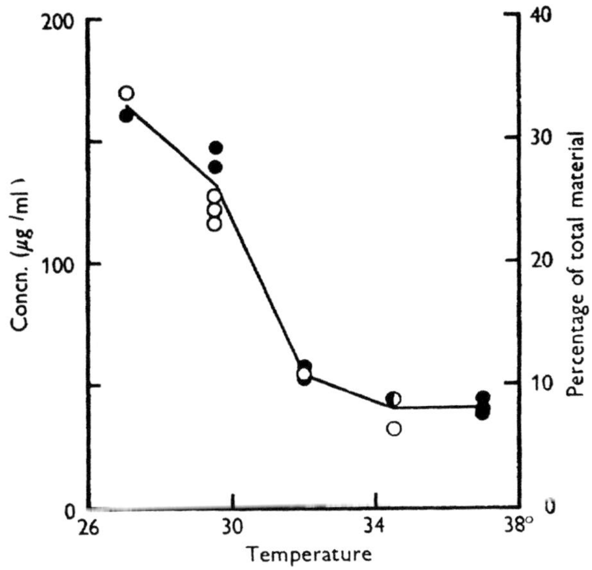

Fig. 3. Effect of temperature on the concentration of collagen in solution. Each sample was initially a solution of 250 µgm of fraction A in 0.5 ml. of 0.067 M phosphate buffer, *pH*=6.8 at 2-4°. ○, results on samples taken directly to the temperature indicated; ●, results on samples first incubated at 37° for 75 min and then taken to the temperature indicated. The left-handed ordinate represents concentration of collagen in solution after incubation for 20 hrs or longer..The right-hand ordinate gives the corresponding percentage of starting material in solution. Each point represents a measurement in duplicate on one sample. From FESSLER (3). Reprinted by permission of the copyright owner, The Biochemical Journal

there is a volume increase of 0.0008 ml/g of collagen when polymerization is carried out in phosphate buffer at *pH* values near 7. Density gradient sedimentation studies in 2 m CsBr were carried out by FESSLER and HODGE (7). The hydrated density of dispersed collagen was found to be less than that of the aggregated form, indicating that water of hydration is lost by collagen molecules passing from solution to fiber. FESSLER (8) estimates that approximately 1/4 of the shell of hydration is lost on polymerization.

Myosin

The skeletal muscles of vertebrates, the striated muscles, are made up of very long, thin multinucleated cells or fibers. The body of the cell is filled with contractile elements, the myofibrils, which form a regular repeating pattern imparting to the cell its characteristic striped or striated appearance. Myofibrils, in turn, are composed principally of two proteins, myosin and actin. Myosin forms thick filaments and actin thin filaments which are interdigitated in the myofibril. Fig. 4 is a diagramatic illustration of the structure of muscle fiber. Interaction between the thick and the thin filaments is responsible for muscle contraction. When the thin filaments and the thick filaments slide together, the muscle contracts; when they slide apart, the muscle relaxes. These protein components of muscle can be extracted and studied in solution.

Pure myosin can be isolated from muscle by appropriate extraction with concentrated salt solutions and subsequently can be purified. The early literature on myosin is somewhat confusing because it has since been shown that what some investigators called myosin is actually a complex of the two proteins, actin and myosin. Nevertheless, there was evidence from the double refraction of flow measurements of von MURALT and EDSALL (9) that rabbit myosin is a highly elongated particle. It is soluble in 3.0 μ buffer, *pH*=7.4, at 0° but not at 23°C (9). WEBER and STÖVER (10) estimated from osmotic pressure studies a molecular weight of about 10^6. While these indications were available as early as 1930, the actual resolution of the size and shape of the myosin molecule required many years. It was recognized in 1955 that the myosin molecule could not be a thread of uniform thickness. It was known at that time that myosin could be degraded by trypsin into two components, light meromyosin and heavy

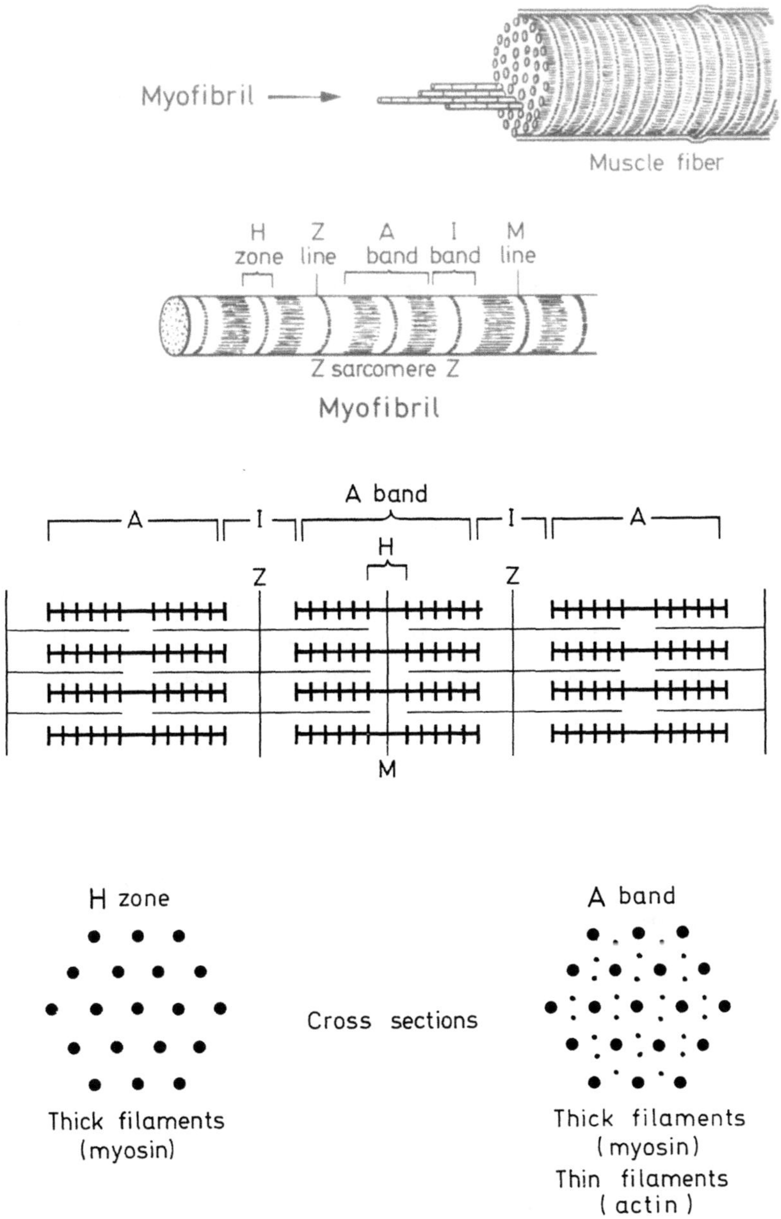

Fig. 4. Schematic representation of the structure of the muscle fiber and the myofibril

meromyosin. The sedimentation coefficient of light meromyosin, s_w^{20}, had been measured and was given as 2.86 x 10^{-13} svedbergs. That of heavy meromyosin had also been measured as 6.98 x 10^{-13}. However, the sedimentation coefficient of myosin was somewhat less than that of heavy meromyosin. LAUFFER and SZENT-GYÖRGYI (11) made a detailed analysis of the sedimentation behavior of non-uniform filaments and concluded that the fact that myosin has a lower sedimentation coefficient than one of its constituents, heavy meromyosin, indicated that myosin was an end-to-end aggregate of very thin, rod-like meromyosin and somewhat thicker rod-like heavy meromyosin. Several models led to the result that the compound could have a lower sedimentation coefficient than one of the components. This idea has been validated, but in somewhat different form, by subsequent studies, reviewed by LOWEY (12). It is now known from measurements of many kinds, including electron microscopy, that myosin is a filamentous molecule about 1500 Å long, consisting of a very thin, long double-stranded fiber with two heads on one end (13), as illustrated diagramatically in Fig. 5. The two heads are in the heavy meromyosin portion of the molecule and the light meromyosin end is simply a thin double-stranded fiber. The overall molecular weight is now regarded as being approximately 500,000.

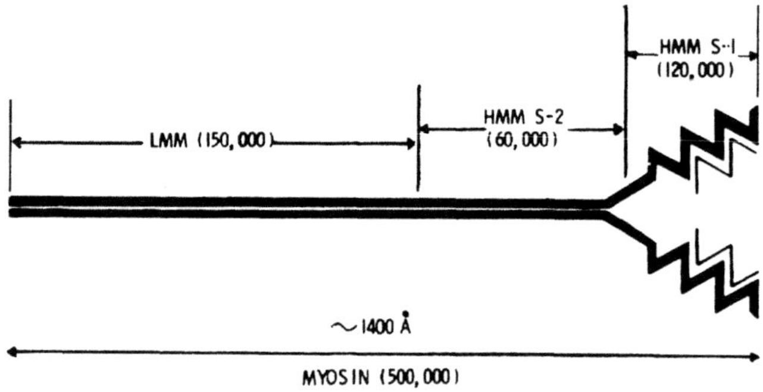

Fig. 5. Schematic representation of the myosin molecule. From LOWEY *et al.* (13). Reprinted by permission of the copyright owner, Academic Press, Inc.

Myosin is a complicated protein. Physically it is differentiated into a gobular head region and a fibrous tail region with high α-helix content. It is not only a structural protein but also an enzyme, possessing ATP-ase activity residing in the two heads. Since in muscle it occurs in thick filaments within the myofibril, it is to be expected that myosin should readily undergo polymerization. The first myosin molecules polymerize tail-to-tail; growth of the fibers proceeds in both directions by tail-to-head association. Fig. 6 is a diagramatic illustration (14). The polymerization of myosin is complex. GODFREY and HARRINGTON (15) found by means of sedimentation equilibrium experiments that myosin, even in salt concentrations as high as 0.5 m KCl, exists as a rapidly reversible equilibrium between monomer and dimer. Since all of the experiments were carried out at 6°C, it is not possible to determine from their studies whether dimerization is an endothermic or an exothermic process. However, in 1955 LAKI and CARROLL (16) reported that the sedimentation coefficient, s^{w}_{20}, of myosin measured at 5°C but not extrapolated to

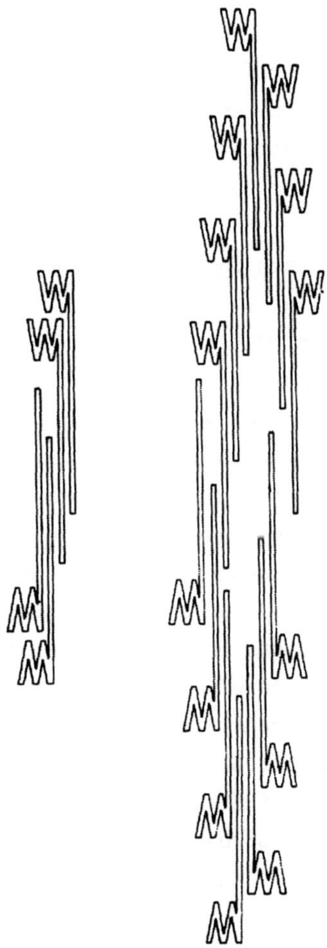

Fig. 6. Schematic diagram of myosin molecules illustrating antiparallel growth of filaments. From "The Mechanism of Muscular Contraction, H.E. HUXLEY (14).

zero concentration is approximately 5×10^{-13}. After standing at room temperature for two hours, s_{20}^{w} increases to 8.2×10^{-13}. This change is reversible, for the sedimentation coefficient returns to approximately 5×10^{-13} when temperature is again reduced. While LAKI and CARROLL did not measure diffusion coefficients at both low and high temperatures and could not, therefore, assert that this reversible change represented aggregation-disaggregation, the possibility is at least suggested by this result that the monomer-dimer transition is endothermic. Based on the analysis of sedimentation behavior made by LAUFFER and SZENT-GYÖRGYI, a change in s_{20}^{w} from 5×10^{-13} to 8.2×10^{-13} could result from side-to-side aggregation of two myosin monomers.

In more dilute electrolyte solutions at *pH* values below 7.3, several boundaries are exhibited in ultracentrifugation experiments, indicating several different kinds of highly polymerized material. Above *pH*=8, however, only two boundaries are observed in the ultracentrifuge, a slow one corresponding to monomer and another corresponding to aggregated myosin with a sedimentation coefficient of 150 S. Electron micrographs of the polymer show that most of the particles have lengths between 5600 and 7500 Å and diameters between 100 and 150 Å. They look very similar to native thick filaments, except that they are shorter (17, 18) JOSEPHS and HARRINGTON (18, 19) estimated the molecular weight of the polymer from viscosity and sedimentation measurements to be in the neighborhood of 50 million. They were able to interpret the results obtained in terms of a rapidly reversible equilibrium between 83 molecules of monomer and 1 molecule of polymer. The equilibrium constant is not affected by temperature; thus one concludes that the enthalpy change is near zero. The equilibrium constant is strongly dependent on *pH* and KCl concentrations. JOSEPHS and HARRINGTON (19) calculated that 11 moles of KCl are liberated and 0.7 moles of hydrogen ion are bound per myosin molecule in forming the polymer. From these facts they calculated a corrected value of the equilibrium constant independent of salt concentration and *pH*. From this, using Eq. (2) of Chap. 1, they calculated the free energy of polymerization at $5^{\circ}C$ to be -2.2 kcal/mole of monomer. Since the enthalpy change is 0, this free energy change of -2.2 kcal/mole at $5^{\circ}C$ corresponds to an entropy increase of 7.8 entropy units per mole of monomer. Thus, this polymerization is entropy-driven.

Pressure within an ultracentrifuge cell varies with the speed of the rotor and with the distance below the surface of the solution. JOSEPHS and HARRINGTON (19) found that at high centrifuge speeds the equilibrium was shifted as the distance below the surface increased. From the change in the equilibrium constant with pressure in experiments carried out under a wide variety of conditions, they calculated that there is a volume increase on the formation of myosin filaments of 0.00065 ± .0001 ml/g of polymerizing myosin.

Even though polymerization of myosin above *pH*=8 is not an endothermic reaction, it is an entropy-driven reaction. Since the equilibrium constant from which the entropy increase was calculated was independent of *pH* and KCl activity, this entropy increase must be independent of the reaction with H^+ ion or KCl. It could well come the release of water molecules.

Actin

The thin filaments in the myofibril are composed of actin, discovered in 1943 by STRAUB (20). Following acetone extraction and removal of myosin from dried muscle powder, actin can be removed by extraction with water. In salt-free solution, actin is a gobular protein, G-actin, with a molecular weight of about 45,000. Upon the addition of neutral salts, G-actin polymerizes into long fibers called F-actin. Details about the polymerization of G-actin have been reviewed by OOSAWA and KASAI (21). The structure of F-actin is illustrated diagramatically in Fig. 7.

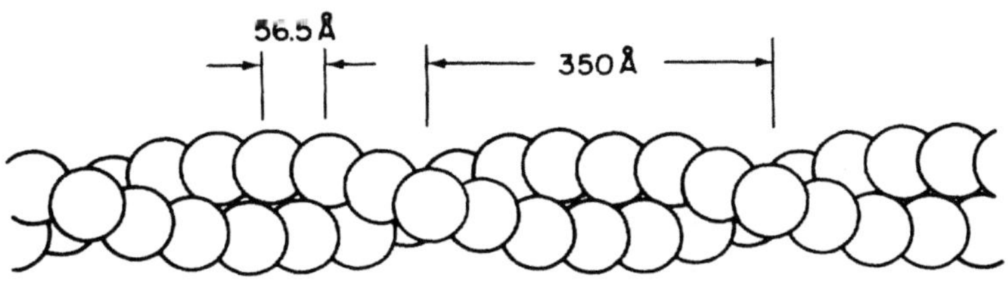

Fig. 7. Diagramatic model of F-actin shown as a two-stranded helix. From OOSAWA and KASAI (21). Reprinted by permission of the copyright owner, Marcel Dekker, Inc.

G-actin binds ATP very strongly. Apparently, one molecule of ATP is bound by each molecule of G-actin. When salt is added to a neutral solution of G-ATP-actin, polymerization and dephosphorylation occur simultaneously, forming F-

ADP-actin (22). When the salt is dialyzed out, F-ADP-actin depolymerizes, but the ADP is not rephosphorylated; the product is G-ADP-actin. This reversible G-F-actin transformation can also be brought about by keeping salt concentration constant and changing temperature. As illustrated in Fig. 8, at an actin concentration of 2.0 mg/ml in a tris-HCl buffer at *pH*=8.0 containing $MgCl_2$ urea and ATP, polymerization occurs at 25°C, and depolymerization at 0°C.

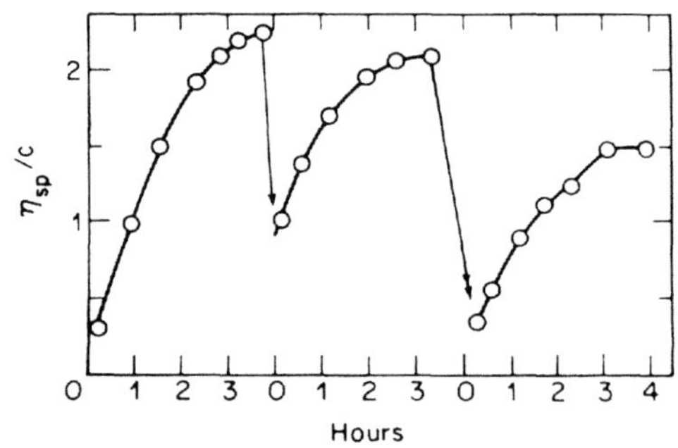

Fig. 8. Reversible G-F transformation of actin by temperature changes. Concentration, 2.0 mg/ml in 5 mM tris-HCl buffer at *pH*=8.0 with 2.0 mM $MgCl_2$, 1.5 M urea and 0.55 mM ATP. Viscosity drops indicated by arrows were induced by incubating the solution at 0°C for 15 and 42 hrs. Temperature was then raised to 25°C, and the viscosity increase was followed. From OOSAWA and KASAI (21). Reprinted by permission of the copyright owner, Marcel Dekker, Inc.

ATP is not necessary for the polymerization. HAYASHI and ROSENBLUTH (23) showed that G-ADP-actin polymerizes to F-ADP-actin in the presence of 0.1 m KCl and 0.001 m $MgCl_2$ at room temperature, but not in the cold. GRANT (24) showed that this polymerization of G-ADP-actin is reversible by changing temperature alone; high temperature favors polymerization, and low temperature favors depolymerization.

Like the other entropy-driven processes we have been considering, this one is accompanied by an increase in volume. This was demonstrated directly by dilatometry by IKKAI *et al.* (25). For the polymerization of G-ATP-actin after correcting for dephosphorylation of ATP, the volume increase was 391 ml/mole. A similar result was obtained for polymerization of G-ADP-actin. It was also shown by IKKAI and OOI (26) that hydrostatic pressure shifts the equilibrium between G and F-actin toward the side of monomer.

It is generally assumed that an equilibrium between monomer and polymer is dynamic. Strictly speaking, the application of thermodynamics to the interpretation of such systems is valid only if this is true. However, it is usually difficult to prove that a dynamic equilibrium in fact exists. The peculiar property of actin, in which ATP is dephosphorylated upon polymerization but ADP is not phosphorylated upon depolymerization, has been used by ASAKURA and OOSAWA, (27) to prove that the equilibrium between monomeric and polymeric actin is dynamic. G-actin binds either ATP or ADP, but it binds ATP more strongly. Thus, the full cycle involves G-ATP actin being transformed to F-ADP actin in the forward direction, and in the backward direction G-ADP actin is formed. Finally, in the presence of ATP, the ADP on the G-actin is displaced by ATP. Thus, if there is indeed a dynamic equilibrium, even though there is no net change in the relative amounts of unpolymerized and poly-

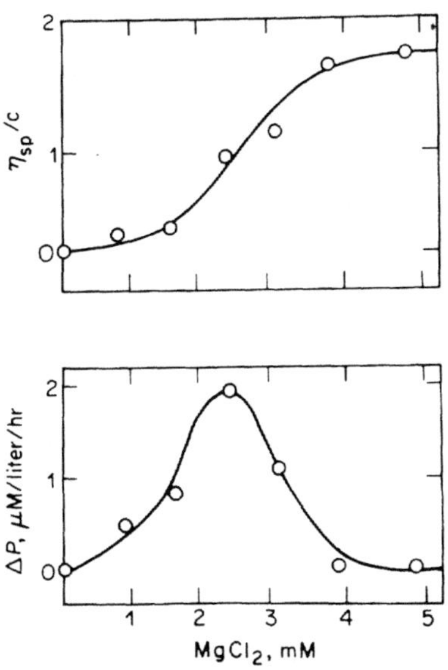

Fig. 9. Reduced viscosity (upper) and rate of ATP splitting (lower) at steady state for actin solutions at various concentrations of $MgCl_2$

Actin concentration, 2.3 mg/ml in 5.5 mM tris-HCl buffer at *pH*=8.2 with 0.6 mM ATP and 1.5 M urea at 20°C. From OOSAWA and KASAI (21). Reprinted by permission of the copyright owner, Marcel Dekker, Inc.

merized actin, there should be a steady rate of dephosphorylation of ATP. ASAKURA and OOSAWA found that this is the case. As illustrated in Fig. 9, at a $MgCl_2$ concentration of 3 mM, where polymerized and unpolymerized actin are both present in a steady state as indicated by an intermediate value for η_{sp}/c, dephosphorylation continues at a high level.

The structure of F-actin, illustrated in Fig. 7, was first deduced by SELBY and BEAR (28) from X-ray diffraction analysis, and it was later confirmed by electron microscopy (29). The F-actin thread is a two-stranded helix composed of gobular monomers with a diameter of about 55 Å. There is a complete twist of the helix every 710 Å; 26 monomeric units are involved in the two strands in a complete twist.

The role of ATP in the polymerization of actin seems to be that of rate regulator. G-ATP-actin polymerizes more readily than G-ADP-actin, but, as has already been seen, G-ADP-actin will polymerize. Indeed, KASAI *et al.* (30) showed that even in the complete absence of nucleotides, actin will polymerize. It is possible to remove ADP from F-actin. This F-actin without nucleotides can be depolymerized, but ordinarily the G-actin obtained is unstable. However, KASAI *et al.* succeeded in stabilizing this G-actin by adding a high concentration of sucrose. The thus stabilized G-actin without nucleotides will polymerize upon the addition of salts.

The polymerization of G-actin to F-actin seems to be a process quite different in its kinetic aspects from the polymerization of TMVP. Available evidence indicates that, in the case of actin, one has an equilibrium between completely unpolymerized G-actin molecules and very long F-actin thread containing several hundred monomers. OOSAWA *et al.* (31), showed that, at intermediate salt concentrations, there is no increase of flow birefringence or viscosity as actin concentration is increased. Above a certain value of the actin concentration, called the critical concentration, C_c, flow birefringence and viscosity increase linerly with actin concentration. This is illustrated in Fig. 10. ASAKURA *et al.* (32) found that the relationships between the total actin concentration and flow birefringence at different temperatures and at different salt concentrations are all expressed by parallel straight lines (Fig. 11). They showed also that the concentration of G-actin remaining behind following removal of F-actin by centrifugation after some of the actin polymerized is a constant equal to C_c, regardless of the total amount of F-actin formed.

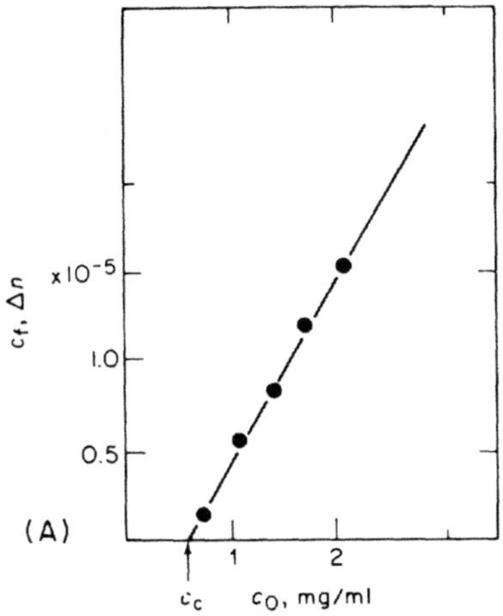

Fig. 10

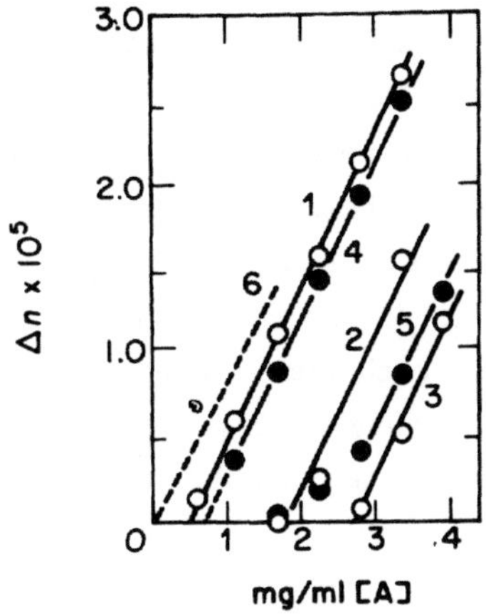

Fig. 11

Fig. 10. The relationship between the extent of flow birefringence and actin concentration at 14°C.

Solvent was 0.8 mM $MgCl_2$ and 8 mM Veronal-HCl buffer at *pH*=8.2. From OOSAWA and KASAI (21). Reprinted by permission of the copyright owner, Marcel Dekker, Inc.

Fig. 11. The temperature effect on the G-F equilibrium of actin solutions at an intermediate salt concentration as followed by the change of flow birefringence, Δn.

Solution contained 0.4 mM $MgCl_2$, 7.5 mM Veronal-HCl buffer at *pH*=8.3 and 0.75 mM ATP. Solutions 1 and 4 were incubated at 20°C; 2 at 6°C; and 3 and 5 at 0°C. Solutions 1, 2 and 3 started from G-actin and 4 and 5 from F-actin. Solution 6 presents completely polymerized F-actin. From OOSAWA and KASAI (21). Reprinted by permission of the copyright owner, Marcel Dekker, Inc.

OOSAWA and KASAI (21) point out that this phenomenon resembles the condensation of gas to liquid or the crystallization of solute molecules from solution. Thus, they regard G-actin polymerization as a condensation phenomenon. This is a different use of the word "condensation" from that used in Chaps. 4 and 5 to describe polymerization of TMVP. In the actin case, "condensation" describes a process resembling condensation of gas molecules to a liquid or the crystallization of a solute, i.e., a phase transition. One might call this a crystallographer's use of the term. In contrast, the polymerization of TMVP can be interpreted, at least over a considerable range of conditions, in terms of the mathematical equations which successfully describe one type of polymerization of synthetic polymers, a type known as condensation polymerization. This is a type of polymerization reaction in which one molecule of water is split out each time a bond is formed in the polymer chain. The unique feature of condensation polymerization is that the length of the polymer chain does not affect measurably the reactivity of the chemical group on the end; it is the same when on the end of a monomer or on the end of a long chain. This fact permits equations to be derived on the basis of the laws of probability, as was done in Chapter 3. The word "condensation" used in connection with TMVP polymerization expresses the fact that these equations have some utility in describing the polymerization of TMVP. In this case, "condensation" is given the definition of polymer chemists. This is just another example of an unfortunate situation in which a single word is used, quite legitimately, to convey radically different meanings.

The kinetics of phase transitions, such as crystallization of a solute, are usually described in terms of at least two processes, the first of which is the formation of nuclei, and the second of which is the growth of these nuclei into crystals. The thermodynamics, however, can be treated much more simply. In a saturated solution, dissolved solute and crystals of the solute are in equilibrium; at the critical concentration, C_c, G-actin is in equilibrium with F-actin. Therefore, the free energy, ΔF, of polymerization is zero. This is the free energy of the transfer of one mole of G-actin to one mole of F-actin when the G-actin is at the critical concentration, C_c. The standard free energy, ΔF°, is the free energy change for the polymerization when the G-actin molecules in solution are at unit activity. The standard free energy, ΔF°, for the polymerization of one mole of G-actin is thus equal to the free energy of the transfer of one mole of G-actin from an activity of one mole per liter to the critical activity. If the solution is nearly ideal at the critical concentration, this is the critical concentration, C_c. ΔF° is given by Eq. (1).

$$\Delta F^{\circ} = RT \ln C_c \quad (1)$$

When combined with Eq. (1) of Chap. 1, this leads to Eq. (2),

$$\ln C_c = \Delta H^{\circ}/RT - \Delta S^{\circ}/R \quad (2)$$

For G-actin, a linear relationship was found between ln C_c and $1/T$ under a wide range of solvent conditions (32). The value of ΔH^o was found to be between 10 and 15 kcal/mole of G-actin. This value was obtained for the polymerization of G-ATP-actin; the reaction is therefore complicated by the dephosphorylation of ATP. However, KASAI (33) found approximately the same value of ΔH^o for the polymerization of G-ADP-actin

The effects of alcohols and acetone on actin polymerization are complex. At room temperature and lower temperatures, these organic solvents favor polymerization. The rate of polymerization in the presence of alcohols increases at low temperatures in proportion to the alcohol concentration. However, at a given alcohol concentration, the rate of polymerization decrease with increasing temperature. At high temperatures alcohol decelerates polymerization (33).

Interaction of Actin and Myosin

The most important reaction of actin is with myosin. This is the basic molecular process involved in muscle contraction. In the absence of ATP, myosin or heavy meromyosin bind only weakly or not at all with G-actin but strongly with F-actin. One heavy meromyosin molecule is bound at saturation to one actin monomer in F-actin. Upon the addition of ATP, the insoluble F-actin-myosin complex shows super precipitation or *in vitro* contraction. This *in vitro* contraction of the actin-myosin complex was found by SZENT-GYÖRGI (34) to be fully reversible thermodynamically. The free energy decrease on contraction was found to be near zero at 0^oC and about 11 kilocalories per mole at 53^oC. Thus, *in vitro* contraction of the actin-myosin complex in the presence of ATP is an entropy-driven process for which values of ΔH^o and ΔS^o can be calculated to be +57 kcal/mole and +207 e.u., respectively.

The interaction of actin and myosin during muscle contraction *in vivo* is mediated by ATP. In the relaxed state, the thick and thin filaments are separated; in the regions of non-overlapping there is no interaction between the myosin of the thick filaments and the actin of the thin filaments. The ATP-ase activity of myosin is centered in the two heads of the myosin molecule at the same site where myosin interacts with actin. ATP is present in the relaxed state. Whenever a nerve impulse is received, membranes in contact with the myofibrils increase their permeability to calcium ions, allowing them to be released. In the presence of calcium ions, ATP is hydrolyzed enzymatically and, by a process which involves tropomyosin, a protein similar to the rod portion of the myosin, and a globular protein, troponin, the heads of the myosin molecules interact with actin subunits to form actomyosin. Details, which are beyond the scope of this book, are the subject of both continuing investigation and extensive speculation. Because of the structure of the myofibril, the only possible way that this interaction can occur is for the thick and the thin filaments to slide together. The more they slide together, the greater will be the number of possible contacts between heads of myosin molecules and actin molecules in the F-actin chains. Muscle relaxation occurs when the ATP is regenerated by the reaction of ADP with phosphocreatine to form creatine and ATP. Many investigators, POLISSAR (35) for example, have postulated that the extent of contraction of a muscle fiber is dependent upon the relative rates of two reactions, one leading to contraction and one leading to elongation, which follow different paths. Thus, the temporary length of a fiber is the result of a steady state, not of a true equilibrium. One way of formulating these two reactions is:

(forward) ATP + myosin + actin $\underline{\underline{Ca^{++}}}$ ADP + actomyosin + phosphate
(reverse) Phosphocreatine + ADP + actomyosin = creatine + ATP + Myosin + + actin.

The question of primary interest in the context of this book is whether an entropy-driven reaction plays a role in muscle contraction in the living organism. It has already been shown that *in vitro* contraction of actomosin is an entropy-driven process. For many reasons, including the complexity of the energy relationships in muscle physiology, it is dangerous to extrapolate directly from this *in vitro* result. However, findings of BROWN (36) suggest strongly that an entropy-driven process is also involved in the living muscle. In the POLISSAR scheme, the isometric tension developed by a muscle is directly proportional to the ratio of the rates of the forward and reverse reactions. BROWN reported that at a variety of pressures up to 6000 psi, the tension developed in the isometric twitch of the auricular muscle of the turtle increases as a linear function of temperature between 2° + 20°C. When the muscle is subjected to pressure at constant temperature, tension decreases with increasing pressure below 5°, remains relatively unchanged between 5 and 8° and increases above $10^{o}C$. This behavior is qualitatively similar to that exhibited by the luciferin-luciferase system described in Chapter 9.

Flagellin

Bacterial organelles of motility, flagella, are normally helical protein filaments with a pitch of about 2.5 μ. Some bacterial strains possess curly flagella with a pitch near 1.2 μ, and still others straight flagella. The studies to be considered here were carried out on normal flagella. Flagella can be removed from bacteria by shaking vigorously, followed by differential centrifugation. When the resulting purified preparations of flagella are suspended in aqueous media and heated at temperatures of about $60^{o}C$, the flagella are depolymerized, ultimately to form monomeric protein units, flagellin, with a molecular weight near 40,000.

Flagellin can be repolymerized to form flagella, provided that an appropriate seed is present. The seed consists of short fragments prepared by ultrasonic degradation of flagella. When such seed material and flagellin are brought together under appropriate conditions (37) the protein monomers add to the end of the short seed fragments to form reconstituted flagella.

The effect of temperature on the reconstitution of flagella from Salamonella strain SJ670 has been reported by GERBER *et al.* (38). The method of investigation was essentially kinetic, involving measurements of the rate of change of the specific viscosity, η_{sp}, when appropriate amounts of seed and flagellin are brought together. The basic data are shown in Fig. 12, where the left ordinate corresponds to the rate of polymerization of flagellin (curve A), and the right ordinate corresponds to the rate of depolymerization of flagella (curve B). The reaction is normally carried out in neutral salt solutions of ionic strength in the neighborhood of 0.1. Curve A shows that the rate of polymerization increases with temperature up to nearly $30^{o}C$ and then decreases abruptly. GERBER *et al.* (38) explained this by postulating that at temperatures above $30^{o}C$ the flagellin monomers capable of polymerizing, M, denature reversibly to a form, M_i, incapable of polymerizing. Earlier studies on the volume increase during polymerization (39) had provided evidence for such a transition. Furthermore, dielectric dispersion data (40) indicate that at temperatures below 26°C, flagellin can be approximated hydrodynamically by an ellipsoid of revolution with axial ratio of 9 to 1, but above 35° by an ellipsoid of ratio 8 to 1.

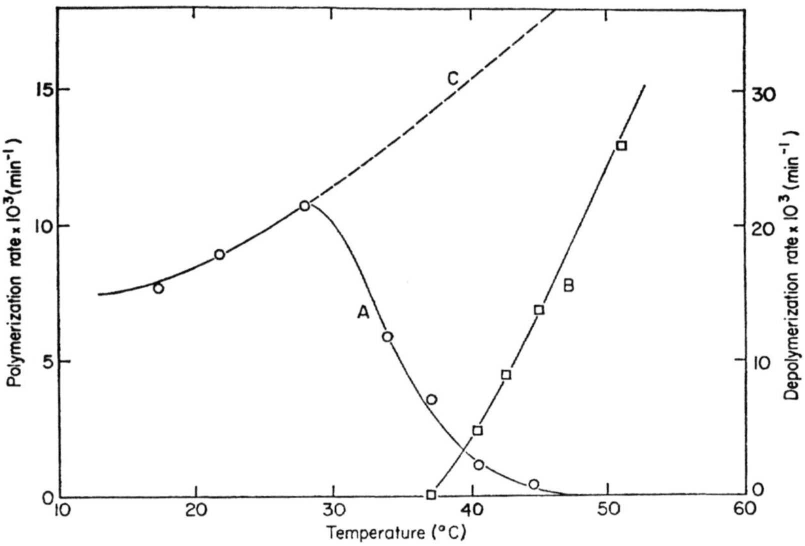

Fig. 12. Effect of temperature on the rates of polymerization or depolymerization of flagellin from Salmonella SJ670. Curve A, rate of polymerization, left ordinate; curve B, rate of deopolymerization, right ordinate. Curve C represents the polymerization rate calculated by extrapolating the low temperature data to higher temperatures in accord with the Arrhenius equation. From GERBER *et al.* (38). Reprinted by permission of the copyright owner, Academic Press, Inc.

GERBER *et al.* (38) estimated the equilibrium constant of this transition by an interesting method. First, they extrapolated the portion of curve A corresponding to temperatures below 30°C, using rate theory (Eq. (8) of Chap. 9) and assuming that the results obtained at the lower temperatures were attributable to pure solutions of the isomer, M. The equilibrium constant, K, for the reaction of M to form M_i is given by Eq. (3).

$$K = [M_i]/[M] \tag{3}$$

GERBER *et al.* then assumed that the rate of change of viscosity with time for polymerization at temperatures above 30°C could be expressed in two ways, as shown in Eq. (4).

$$d\eta_{sp}/dt = k_m([M] + [M_i]) = k_c[M] \tag{4}$$

The rate constant, k_m, is the one taken directly from curve A, and it must be multiplied by the total protein concentration $[M] + [M_i]$. In contrast, k_c is the value taken from curve C at the same temperature and would be appropriately multiplied by [M]. By combining Eqs. (3) and (4) one obtains Eq. (5).

$$K = (k_c/k_m)-1 \tag{5}$$

In this manner, K was calculated at various temperatures above 30°C. When the logarithms of such K values were plotted against $1/T$ in the manner of Eq. (3) of Chap. 1, the data fell on a straight line from which values of ΔH^o and ΔS^o of +101,000 cal/mole and +332 e.u. were obtained. By comparing these results with others obtained by different methods in their own laboratory, GERBER *et al.* concluded that the average and most probable values of ΔH^o and ΔS^o should be 110,000 cal/mole and 362 e.u., respectively. It must be remembered that this result applies to the postulated equilibrium for denaturation of monomer.

Information about the postulated equilibrium between monomer and flagella can be obtained by focusing attention on the temperature range between 37°C and about 45°C in Fig.12, the region in which both polymerization and depolymerization take place at measurable rates. GERBER *et al.* (38) made the assumption that at equilibrium, specific viscosity would not change with time; $d\eta_{sp}/dt$ is zero. They therefore made up solutions containing different amounts of seed and different amounts of monomer and measured the rate of change of specific viscosity for each and extrapolated the results to zero rate of change. In this way, they found that, at constant temperature, regardless of the concentration of seed or the initial concentration of monomer, the concentration of monomer, $[M_c]$, corresponding to equilibrium was the same within rather small experimental error. This indirect method, therefore, is a way of estimating critical concentration of monomer at each temperature studied in equilibrium with any amount of polymer. Eq. (2) can therefore be used to evaluate the thermodynamic parameters. When ln $[M_c]$ was plotted against $1/T$, the results shown in Fig.13 were obtained, from which the value of the enthalpy, ΔH_c^o, was found to be -70,000 cal. This appears therefore to be an exothermic reaction.

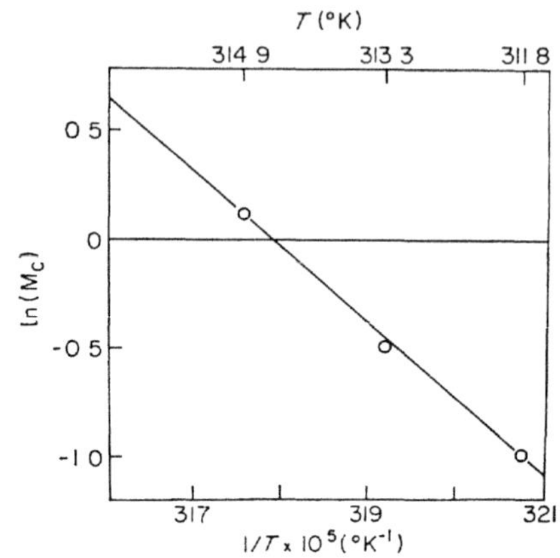

Fig. 13. The relationship between the equilibrium monomer concentration, M_c, and temperature for Salmonella SJ670 flagellin. From GERBER *et al.* (38). Reprinted by permission of the copyright owner, Academic Press, Inc.

However, according to the postulates of GERBER *et al.* (38), the critical concentration is the sum of the concentration of unpolymerizable, denatured, protein, $[M_i]$, and the concentration of polymerized protein, [M], as shown in Eq. (6).

$$[M_c] = [M_i] + [M] = K[M] + [M] = (K + 1)[M] \simeq K\,[M] \tag{6}$$

The final approximation is very accurate because, at temperatures where both polymerization and depolymerization can be measured, the value of K for denaturation is enormous. Eq. (7) and (8) follow directly from Eq. (6).

$$d\ln[M]/d(1/T) = d\ln[M_c]/d(1/T) - d\ln K/d(1/T) \tag{7}$$

$$-\frac{\Delta H^{\circ\prime}}{R} = \frac{70{,}000}{R} - \frac{110{,}000}{R} = -\frac{40{,}000}{R} \tag{8}$$

Therefore, by this method of reasoning, $\Delta H^{\circ\prime}$, the enthalpy of polymerization of the polymerizable fraction of monomers is +40,000 cal/mole. It is concluded therefore that the actual polymerization step is endothermic and, therefore, of necessity entropy-driven.

This case illustrates the fact that some reactions of concern which seem at first glance to be exothermic might actually be endothermic or entropy-driven when analyzed in greater detail. The approach to this problem is interesting in that it shows how indirect methods can sometimes be substituted for direct thermodynamic measurements. It is necessary, however, to recognize that the analysis is strongly dependent upon extrapolation of data, always a process subject to some risk, and that the assumed equilibrium between monomer and polymer was not confirmed by thermodynamic methods.

Summary

The three important proteins in the animal body, collagen, actin and myosin, all of which perform their function in a fibrous state, are formed *in vitro* by entropy-driven polymerization of smaller units. Even though the process by which the fibrous forms of these proteins are produced in the living cell might be more complex than in the isolated condition, nevertheless, since the initial and final states are at least similar *in vitro* and *in vivo*, contributions to the negative free energy of polymerization identified by experiments on purified extracts must also be present in the living cell. Since it has been established that the entropy increase necessary to drive the polymerization of tobacco mosaic virus protein comes from the release of water molecules, it is very probable that the same source accounts for the increase in entropy in these entropy-driven reactions. If this is so, one is forced to the conclusion that water plays an enormously important role in the generation of structures which play the dominant role in such important body tissues as tendon and muscle.

It is tempting to speculate that water interaction might also play a role in muscle contraction. This is suggested by the fact that *in vitro* contraction of acto-myosin preparations is an endothermic process. While the exact relationship between *in vitro* contraction and muscle contraction in the living system is not clearly understood, studies on the effect of temperature and pressure on the isometric tension developed in the auricular muscle of the turtle support the tentative conclusion that muscle contraction involves a similar process.

The protein flagellin extracted from normal flagella of bacteria, in the presence of appropriate seed, can be polymerized to form flagella. GERBER postulates that, at temperatures above 30°C, the flagellin monomers capable of polymerizing are in equilibrium with a denatured form, incapable of polymerizing. The polymerization of flagellin, consisting of both the polymerizable native form and the unpolymerizable denatured form in equilibrium with it, is an exothermic process at temperatures above 30°C. However, when correction is made for the enthalpy of the transition from the denatured form, present in great excess, to the native form, the polymerization of the native forms turns out to be endothermic. This reaction is, therefore, entropy-driven.

Chapter 9

Other Entropy-Driven Reactions Involving Macromolecules

Gels

The first entropy-driven synthetic processes discovered involved the transformation of sols into gels. In 1924 SZEGVARI (1) observed that when nitrocellulose is dissolved in amyl acetate, ether-alcohol or acetone and then diluted with solvents which do not dissolve nitrocellulose, such as benzine, benzene, tetralin, etc., a viscous sol is formed at room temperature. Upon raising the temperature, one obtains a rigid gel, which again becomes fluid when the temperature is lowered. SZEGVARI explained this in terms of competition between the nitrocellulose and the nonsolvent (benzine, etc.) for the solvent (amyl acetate etc.). At high temperature, the affinity of the solvent for the nonsolvent out-weighs the affinity of the solvent for nitrocellulose; a gel is formed. At lower temperatures, a sol is formed because the affinities are reversed.

Soon after this, McBAIN *et al.* (2) obtained somewhat similar results with nitro-cotton dissolved in anhydrous alcohol. A solution at 0.8% of nitro-cotton, reacted to the extent of 12.1% nitrogen, dissolved in anhydrous alcohol formed a viscous sol at the temperature of dry ice-ether. This became a flowing jelly or gel at room temperature. A solution containing 5% nitro-cotton was a stiff, very turbid gel. All of these gels became sols again on cooling to the dry ice-ether temperature.

HEYMAN (3) pointed out that there are three types of gel, differing with respect to the energetics of the sol-gel transformation. In one type, a sol-gel transformation is accompanied by an evolution of heat and a small shrinkage of volume. The prime example of such is gelatin. In the second type, there is very little heat change and no volume change in the transformation.

Such gels are thixotropic, as exemplified by vanadium pentoxide. The third type exhibits an absorption of heat and a small increase in volume upon gelation. This is obviously an entropy-driven gelation; the two systems discussed above fit this category.

HEYMAN carried out studies on the gelation of methylcellulose. He reported that colloidal aqueous solutions of methylcellulose show an increase in volume upon gelation, and that gel formation is favored at higher temperatures and liquefaction at lower temperatures. FREUNDLICH (4) discussed these experiments and calculated from HEYMAN's dilatometric measurements that the volume increase is between 0.08 and 0.13 ml/100g upon gelation. HEYMAN and FREUNDLICH both explained this change by assuming that in the gel state the methylcellulose particles are less hydrated than in the sol state. Thus, the idea that endothermic polymerization in aqueous solution involves freeing of water molecules is very old.

MARSLAND and BROWN (5) carried out further studies on the gelation of methylcellulose. They did their research with a pressure bomb in which a chrome steel ball was allowed to fall under gravity through a sample of the gel. The strength of the gel was considered to be proportional to the time it took the steel ball to fall from an upper window to a lower window after being set in motion by quickly rotating the bomb to a vertical position. Measurements were made at various pressures ranging from atmospheric up to 8,000 pounds per square inch (psi). The bomb was immersed in a water bath, the temperature of which could be controlled to within $\pm 0.1^{\circ}C$. The results are shown in Fig. 1, where $\log (N_1/N) \times 100$ is plotted against pressure. N_1 is the gel strength at a given pressure and N is that at atmospheric pressure, both measured as indicated above. These results were interpreted in terms of Eq. (17) of Chap. 1 to calculate $\Delta\overline{V}$, the increase in volume per mole upon gelation. This equation relates $\ln K/K_0$ to $\Delta\overline{V}$, where K and K_0 mean the equilibrium constant for the gelation process at a given pressure at atmospheric pressure, respectively. It was assumed that K/K_0 is equal to N_1/N.

In a discussion of experiments similar to this, JOHNSON, EYRING and POLISSAR (6) derived this relationship on the assumption that the gel constituents exist in a folded, globular, comparatively nonviscous state at a concentration, C_g, in equilibrium with an unfolded, fibrous state at concentration, C_f. Eq. (1) is the appropriate expression for the equilibrium constant when the molecular weights of the globular and the fibrous particles are the same.

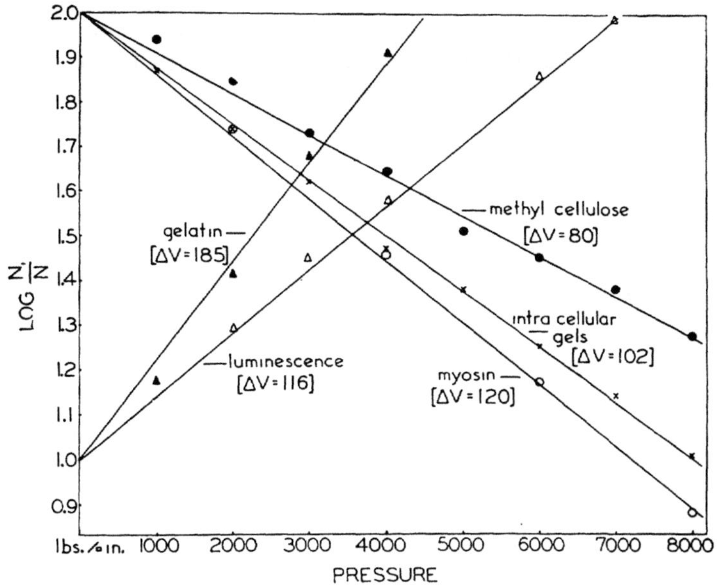

Fig. 1. Effects of pressure upon gels. N_1 = various values of gel strength at different pressures (in pounds per square inch, and N = values at atmospheric pressure. ΔV is the cubic centimeter per mole change in volume which the substance undergoes in passing from the sol to the gel state, or vice versa. Myosin (from rabbit muscle): 2.1% solution, 36.7°C., *pH*=6.72. Methyl cellulose: 1.66% solution, 54.0°C. Gelatin: 4% solution, 22.0°C., *pH*=6.4 Luminescence curve shows data for reactivation of *P*, phosphoreum luminescence, at 34°C. Intracellular gels: Average curve for the gelation values of the plasmal gel system of Amoeba (two species), Arbacia eggs (fertilized and unfertilized), Elodea (leaf cells), and also for the streaming rate in Elodea and the rate of furrowing in Arbacia eggs. From MARSLAND and BROWN (5). Reprinted by permission of the copyright owner, the Rockefeller University Press

$$K = C_f/C_g \tag{1}$$

JOHNSON *et al.* then assumed that the viscosity, η, of the solution is given by Eq. (2).

$$\eta = \eta_0 + aC_f + bC_g \simeq aC_f, \tag{2}$$

where a and b are constants. In their formulation, C_0 equals $C_g + C_f$. They set the maximum viscosity, η_m equal to aC_0. One can then write Eq. (3).

$$K = C_f/C_g = \frac{\eta}{\eta_m - \eta} = \frac{1}{(\eta_m/\eta) - 1} \tag{3}$$

When η_m/η is much greater than one, $K = \eta/\eta_m$. Therefore,

$$K/K_0 = \eta_1/\eta = N_1/N. \tag{4}$$

Thus Eq. (17) of Chap. 1 can be written,

$$\log N_1/N = -(P - P_0)\ \Delta\bar{V}/2.3\ RT. \tag{5}$$

When plotted in the manner shown in Fig. 1, a straight line should result.

The value of $\Delta\bar{V}$ calculated from these data in this manner was reported by MARSLAND and BROWN to equal +80 ml/mole. FREUNDLICH estimated from HEYMAN's dilatometric data that the volume increase was between 0.08 and 0.13 ml/100 g. One can infer, therefore, that the molecular weight of the methylcellulose undergoing gelation is of the order of magnitude of 10^5.

As is shown in Fig. 1, MARSLAND and BROWN also studied the sol-gel transformation of rabbit-muscle myosin. Here, too, hydrostatic pressure shifted the equilibrium toward sol. From the results presented in Fig. 1, using the theory presented above, they calculated that the increase in volume per mole of myosin upon gelation is 120 ml. Initially, myosin forms a gel at 37°C which becomes a sol either on compression or cooling. However, the myosin undergoes thermal denaturation and sets into a very firm gel which cannot be reversed.

Also shown in Fig. 1 are results on intracellular gels. The relationship of intracellular gels to such biological processes as cell division, formation of the spindle apparatus etc. will be described in Chap. 10. The strength of these gels was measured by determining the time for a standard amount of movement of pigment granules contained within them. It will be shown that such gels probably are formed as the result of polymerization of a protein, tubulin, into microtubules. This being the case, the model used by MARSLAND and BROWN for the interpretation of the gelation of methylcellulose and of myosin seems inappropriate for the interpretation of the gel strength of the intracellular gel. It would be more appropriate to assume that gel strength is proportional to the square of the weight average molecular weight of the polymerized material[1], and that the polymerization follows the mathematics of condensation polymerization. If Eqs. (2) and (4) of Chap. 4 are combined, the result is,

$$\tau/\tau_0 = \sum_1^\infty \frac{Hc_x M_x}{Hc_0 M_0} = \frac{1}{M_0} \sum_1^\infty \frac{c_x M_x}{c_0} \equiv \frac{\overline{M}_w}{M_0} \tag{6}$$

Thus, $\overline{M}_w/M_0$ can be substituted into Eq. (6) of Chap. 4 to obtain,

$$K = \frac{(\overline{M}_w/M_0)^2 - 1}{4[\mathrm{P}]_0} \simeq \overline{M}_w^{\,2}/4M_0^2[\mathrm{P}]_0 , \tag{7}$$

when $\overline{M}_w \gg M_0$. Therefore, N_1/N can be substituted for K/K_0 as was done in Eq. (4). MARSLAND and BROWN estimated ΔV to be 102 ml/mole when they interpreted their results in terms of Eq. (5). The best estimate is that tubulin is a dimer with a molecular weight of approximately 120,000. Thus, the volume expansion would be the equivalent of 0.00085 ml/g. This must be compared with the volume increase obtained by dilatometry for the polymerization of TMVP, 0.0079 ml/g.

Soaps and Synthetic Detergents

An endothermic sol-gel transformation is exhibited by the soap, sodium behenate (7). On rapid cooling to room temperature, solutions form a limpid mobile liquid. Reheating this solution to about 90°C produces a clear, stiff isotropic jelly, which quick cooling once more renders fluid. However, the fluid at room temperature is not stable and precipitates in a short time. Slow cooling of the 90° gel produces a hard, white curd.

[1]This is a crude approximation because viscosity (the measure of gel strenght) is proportional to squared molecular weight only for long rods and because the viscosity of an inhomogeneous system is proportional to the weight average of the molecular weight squared - a quantity different from the square of weight average molecular weight.

Synthetic detergents and soaps form micelles in aqueous solution. DODDARD *et al.* (8) measured calorimetrically heats of micelle formation at 25°C at concentrations near the critical micelle concentration. Positive values of 0.5, 0.8 and 0.3 kcal/mole, respectively, were obtained for sodium decyl sulfate, sodium octyl sulfate and sodium octyl sulfate in 0.69 m NaCl. Similar measurements were made on the soap, potassium octanoate in 0.42 m KOH and a value of 1.7 kcal/mole was obtained.

Antibodies

The thermodynamics of the reaction between rabbit antiarsanilic antibody and two haptens, terephthalanilide-pp'diarsonic acid and adiptanilide-pp'diarsonic acid was studied by DOTY and EPSTEIN (9). The standard free energy change at 25°C, the enthalpy change and the standard entropy change for the formation of complex in 0.15 ionic strength borate buffer, *pH*=8, were found to be -7.5 kcal/mole, -2.4 kcal/mole and +17 e.u., respectively. It is evident, therefore, that the major fraction of the driving force for this reaction comes from the increase of entropy; it is an entropic process.

HAUROWITZ *et al.* (10) observed that antigen-antibody precipitates obtained in the region of antibody excess dissociate on washing with 0.9% saline solution into free antibody and an insoluble antigen-antibody complex. As antigens, they used three proteins, beef serum pseudoglobulin, beef serum gamma globulin and crystalline egg albumin, each coupled with arsanlic, anthranilic or sulfanilic acid. Appropriate antibodies were prepared. Using isotopically labelled radioactive antigen or antibody, they found that the saline extracts of the antigen antibody precipitates contained antibody but no antigen. More antibody could be extracted from the precipitates at 25°C than at 3°C. From these studies they were able to calculate the standard free energy for the combination of one mole of antibody with antigen antibody complex. This was found to vary between -8 and -10 kcal/mole of antibody. The enthalpy values varied from about -3.2 to 7.5 kcal. Thus, in all cases a significant fraction of the total negative free energy for antibody binding comes from positive entropy changes. This reaction is, therefore, at least partially entropy-driven.

Proteins and Synthetic Polypeptides

SCOTT and BERNS (11) studied the polymerization of phycocyanin, a chromoprotein from blue-green algae with a monomeric molecular weight of approximately 30,000. They studied the system by carrying out sedimentation velocity experiments as a function of *pH*, ionic strength, temperature and buffer. They interpreted their data in terms of an equilibrium between monomer, trimer, hexamer and dodecamer. From the variation of the equilibrium constant between trimer and hexamer with temperature, they estimated that the enthalpy of this transition is 24 kcal/mole and the standard entropy is +100 entropy units. Thus, this particular transition is endothermic and entropy-driven. In contrast, the polymerization of hexamer into dodecamer was found to be slightly exothermic with a standard entropy of -26 entropy units.

NOGUCHI *et al.* (12) found that water-soluble poly-ε-aminocaproyl-DL-alanine with a molecular weight of 16,700 polymerized reversibly with increasing temperature. This polymerization is *pH* sensitive; at a polymer concentration of 0.4% it takes place in the neighborhood of 45°C at *pH*=4.0 and near 65°C at *pH*=9.0. At *pH*=10.4, there is no polymerization at any temperature. The enthalpy of polymerization is therefore positive and the process must of necessity be entropy-driven.

Poly-L-proline exists in two forms; (I) is a right-handed helix and it is insoluble in water; (II) is a left-handed helix and is water-soluble. The effect of temperature on the polymerization of form (II) was studied in some detail by CIFERRI and ORIFINO (13). Poly-L-proline (II) was brought into aqueous solutions by lowering the temperature to the range, -15°C to $+5^{o}$C. When a 0.4% solution in pure water was heated at the rate of 1° per day, starting at 50°C, turbidity and a precipitate were visually observed to form at 54°C. Different appearance temperatures were observed in various salt solutions. The formation of precipitate could be arrested by dropping the temperature to 25°C, but complete solublization occurred only on dropping the temperature to the range, $+5^{o}$ to -15°C. BROWN and JENNINGS (14) obtained more or less similar results by making electric birefringence (Kerr electro-optic effect) measurements. Their results indicated that even at low temperature, the material is aggregated to a considerable extent, but less than at high temperatures.

The helix coil transition of poly-γ-benzyl-L-glutamate in ethylene dichloride-dichloro acetic acid mixtures was studied by DOTY and YANG (15). They observed an increase in $[\alpha]_D$ with temperature. Thus, the enthalpy of the transition from coil to helix must be positive, and since the transition is spontaneous there must be a positive entropy change, even though internal order is being created. Thus, this transition constitutes another example of an entropy-driven process.

SCHADE and REINHART isolated a protein from fresh human milk at neutral *pH* to which they gave the name galactothermin (16). This protein, which has a molecular weight of around 14,000, aggregates reversibly. Increasing concentration, ionic strength and temperature and approach to the isoionic *pH*, 7.27, all favor aggregation. MURAYAMA (17) reported an enthalpy of activation of +70 kcal/mole and an entropy of activation of +220 e.u. High pressure studies indicate that ΔV of activation is +70 ml/mole. These parameters were determined from rate experiments and their significance is explained in the section of this Chapter on thermal denaturation of proteins.

Poly-L-Valyl-Ribonuclease, containing a total of 22 valine residues distributed over seven reactive sites, undergoes thermal aggregation but untreated ribonuclease does not (18). KETTMAN *et al.*, (18) showed that at a concentration of 2 mg/ml in 0.5 M phosphate buffer at *pH*=6.8, this material aggregates at 30° and also at 39°, but dissociates at 0-4°C. Aggregation is promoted by increasing ionic strength. Pressure studies indicated a volume increase of approximately 200 ml/mole during aggregation.

The polymerization of insulin has been studied extensively. ONCLEY *et al.* (19) found that the sedimentation coefficient of insulin depended on concentration, ionic strength and *pH*. The minimum sedimentation coefficient was assumed to correspond to a monomeric particle with a molecular weight of about 12,000 and higher coefficients were attributed to dimers, trimers, tetramers, etc. and equilibrium mixtures of these. WAUGH *et al.* (20, 21, 22) showed that insulin, upon heating, polymerizes to form fibers and ultimately precipitates.

The rate of coagulation has a temperature coefficient, Q_{10}, of about 4.0 in sulfuric acid. This might be taken as presumptive evidence that fiber formation is an endothermic reaction. However, even though monomeric insulin can be obtained from fibers by indirect methods, no evidence was presented for a strict temperature-dependent equilibrium between unpolymerized and polymerized insulin. It is therefore not possible to state definitely that fibril formation in insulin is an entropy-driven reaction.

There is, however, evidence that early stages of insulin polymerization are partially entropy-driven. DOTY and MYERS (23) investigated by light scatter measurements the transition between monomeric insulin with molecular weight 12,000 and dimers, trimers, etc. These transitions are temperature sensitive. From the variation of equilibrium constants with temperature, DOTY and MYERS evaluated the enthalpy and entropy of dissociation. The entropy of dissociation turned out to be positive, but the average value was only six e.u. With the aid of the Sackur-Tetrode equation, as discussed in Chap. 5, they estimated that the entropy change attributable to translation and rotation for the dissociation of one mole of dimer into two moles of monomer should be approximately +122 e.u. instead of 6 as observed. They postulated that the low value observed indicated that water molecules are bound when dimer dissociates into monomer. Thus, dimerization involves the release of water molecules with an increase in entropy. There is therefore an entropic contribution to dimerization, but other interactions involving a decrease in enthalpy must also be involved.

As with fiber formation by insulin, the polymerization of nucleohistones is strongly dependent upon temperature (14, 25). Aggregation is favored by raising *pH* and ionic strength. The rate of aggregation at fixed *pH* and ionic strength accelerates sharply with increase in temperature. However, no evidence is presented for a temperature-dependent equilibrium between polymerized and unpolymerized forms.

Ion Binding by Proteins

Numerous examples, reviewed by KLOTZ (26), of entropy-driven reactions between proteins and ions are known. From the free-energy values for binding at $0^{o}C$ and $25^{o}C$, KLOTZ calculated the enthalpy and entropy of binding of copper by serum albumin. For the first through the sixteenth copper ions bound, he found values for the enthalpy varying from +2.8 kcal/mole to +3.7 kcal/mole and for the entropy values varying between 29.2 e.u. and 18.0 e.u., respectively. Positive entropy changes were also found for the binding of anions by serum albumin. The entropy of binding was positive for chloride, octyl sulfate, decyl sulfate, dodecyl sulfate, methyl orange and azosulfathiazole. The enthalpy of binding was positive in some cases and negative in others. KLOTZ assumed that the release of water molecules was the source of the positive entropy of binding.

Precipitation of Proteins

COHN and FERRY (27) report positive heats of precipitation (negative heats of solution) for egg albumin in 2.14 M ammonium sulfate between 0 and 12°C, and for carboxyhemoglobin in 1.267 M phosphate buffer at *pH*=6.6 between 0 and 25°C, and for horse serum albumin sulfate in water at temperature between 5 and 25°C. ANDERER and LAUFFER found positive heats of crystallization for tobacco mosaic virus (28). WARNER found that β-casein from bovine milk, at *pH*=4.9, is soluble at 2°C, but insoluble at room temperature (29).

Thermal Denaturation of Proteins

As a general rule, the thermal denaturation of a protein is accompanied by a very high enthalpy of activation, frequently between 50,000 and 100,000 cal/mole. Such enthalpies of activation are obtained by interpreting the variation of the rate constant for denaturation with the reciprocal of the absolute temperature through the use of the Arrhenius equation or its modern equivalent, the theory of absolute reaction rates, a process almost analogous to the evaluation of van't Hoff enthalpies. According to the theory of absolute reaction rates (30), an intermediate step in denaturation is the transition of a native molecule into a hypothetical critical reactive form, the activated complex; the enthalpy and entropy of activation characterize this process. The rate of a reaction is assumed to depend upon the fraction of molecules in the critical form and upon the probability that a molecule in that state will undergo chemical change. The theory can be expressed by Eq. (8).

$$k = \frac{\kappa\, RT}{Av\, h}\, e^{\Delta S^*/R}\, e^{-\Delta H^*/RT} \qquad (8)$$

In Eq. (8), k is the specific reaction rate, κ is a transmission coefficient usually assumed to be unity for protein reactions, R is the gas constant, T is the absolute temperature, Av is Avagadro's constant, h is Planck's constant and ΔS^* and ΔH^* are the entropy and enthalpy, respectively, of the transformation of a molecule into the critical reactive form, the entropy and enthalpy of activation. The second step is the transformation of the critical reactive form into the denatured form. This is true whether the denaturation is carried out reversibly or irreversibly.

While thermal denaturation studies are frequently carried out on irreversible processes, with some proteins reversible denaturation can be achieved. One example, the reversible transformation of polymerizable flagellin into a unpolymerizable form, has already been discussed in Chap. 8. The van't Hoff enthalpy and the entropy of this process are +110,000 cal/mole and +362 e.u., respectively. Other examples of proteins capable of being denatured reversibly are hemoglobin, serum albumin, trypsin, chymotrypsin, trypsinogen, chymotrypsinogen and pepsinogen (31).

Both the kinetics and the thermodynamics of the reversible denaturation of crystalline soybean trypsin inhibitor have been studied extensively by KUNITZ (32). The findings with this protein are presented as an example. Crystalline soybean trypsin inhibitor is denatured when it is heated in 0.0025 M HCl at a protein concentration of 0.5%. Such solution have a pH of about 3. When the heated protein is poured into a larger volume of very cold 0.06 M acetate buffer at pH=4.5, the denatured fraction precipitates and the native protein remains in solution. When one measures the amount of protein remaining uncoagulated as a function of time of heating at different temperatures, one can evaluate the enthalpy of activation. However, if the protein in the hydrochloric acid solution is cooled to about 25°C and allowed to stand for several days, the reaction is reversed and the native form of the protein is restored. KUNITZ demonstrated this by showing that after storage at 25°C for sufficient time, the protein no longer coagulated when mixed with cold acetate solution. Thus, by measuring the maximum extent of denaturation at a particular temperature, one can obtain values of the equilibrium concentrations of native and denatured protein. This permits evaluation of the equilibrium constant for denaturation and, from its variation with temperature, van't Hoff enthalpy and entropy can be obtained.

In the manner indicated, KUNITZ obtained an enthalpy of activation and an entropy of activation of +55,350 cal/mole and +95 e.u., respectively. The van't Hoff enthalpy of reversible denaturation, ΔH^{o}, was found to be +57,300 cal/mole and the entropy, ΔS^{o}, was found to be +180 entropy units. Reversible denaturation of crystalline soybean trypsin inhibitor, like the denaturation of flagellin, is clearly an entropy-driven reaction. In the reversible case, renaturation is also assumed to pass through the critical reactive form. It is a thermodynamic necessity that the van't Hoff enthalpy of the equilibrium process must equal the enthalpy of activation of the forward process minus the enthalpy of activation for the reverse process. It follows that the same must be true for the entropy change. Since the van't Hoff enthalpy of the equilibrium reaction is +57,300 cal/mole and the enthalpy of activation of the forward process is +55,300 cal/mole, the enthalpy of activation for the reverse process must be -1900 cal/mole. Again, from the entropy of the reversible reaction, +180 e.u., and the entropy of activation of the forward process, +95 e.u., the entropy of activation of the reverse process must be about -85 e.u. In this specific example, the van't Hoff enthalpy of the reversible reaction and the enthalpy of activation of the forward reaction are both large positive numbers. Thus, the high positive enthalpy of activation for denaturation correlates with the fact that denaturation is an entropy-driven process.

In Chap. 8 it was shown that the reversible denaturation of flagellin exercised control of the reversible polymerization of the protein. A previously investigated example of a biological process controlled by reversible entropy-driven protein denaturation is afforded by the bioluminescence studies of BROWN *et al.* (33). Luminescent organisms include the fire fly and numerous crustaceans, worms, fungi, protozoa and bacteria. The general explanation of bioluminescence is that light is emitted when a heterocyclic phenol, luciferin, is oxidized. A heat-labile enzyme, luciferase, catalyzes the reaction. The first step is the formation of a luciferin-AMP-enzyme complex by the reaction of luciferin, ATP and luciferase in the presence of magnesium ion. This enzyme complex is then oxidized by molecular oxygen to yield products in an excited state which subsequently emit light.

The brightness of the luminescence in *Photobacterium phosphoreum* was measured by BROWN *et al.* (33) at various temperatures and pressures. The results are shown in Fig. 2. It can be seen that there is an optimum temperature for the brightness of luminescence in the neighborhood of 20 - 25°C, and that the variation of the brightness with temperature depends upon pressure. These data have been interpreted quantitatively by EYRING and MAGEE (34) on the basis of the assumption that the emission of light is limited by the activity of luciferin and that there is an equilibrium between active luciferase and an inactive or denatured form. By analogy with Eq. (6) of Chap. 8, one can write Eq. (9), where $[A_n]$ and $[A_o]$ mean the concentration of the native or active form of the enzyme and the total enzyme concentration, respectively, and K is the equilibrium constant for denaturation.

$$[A_n] = [A_o]/(1 + K) \qquad (9)$$

EYRING and MAGEE assumed that in the presence of excess luciferin, L, the intensity of luminescence, I, is given by Eq. (10).

$$I = bk[L][A_n] = bk[L][A_o]/(1 + K) \qquad (10)$$

In this equation, k is the rate constant for the formation of enzyme substrate complex, and b is a proportionality constant. When the expression for K from Eq. (18) of Chap. 1 and, from Eq. (8), the expression for k obtained from the theory of absolute reaction rates are substituted into Eq. (10) and when all constants are incorporated into c, Eq. (11) results.

$$I = \frac{cTe^{-\Delta H_o^*/RT}\, e^{-(P-P_o)\,\Delta\overline{V}^*/RT}\, e^{\Delta S_o^*/R}}{1 + e^{-\Delta H_o^o/RT}\, e^{-(P-P_o)\Delta\overline{V}/RT}\, e^{\Delta S_o^o/R}} \qquad (11)$$

In this equation, the asterisks identify the parameters appropriate for the activation step in the theory of absolute reaction rates, and the subscript o identifies values of parameters at atmospheric pressure. The curves in Fig. 2 were calculated with the aid of Eq. (11) when the following values were assigned to the parameter; $\Delta H_o^* = 17{,}220$ and $\Delta H_o^o = 55{,}260$ cal/mole, $\Delta\overline{V}^* = 50$ at 0^oC and $\Delta\overline{V} = 64.6$ ml/mole at 35^oC and $\Delta S_o^o = 184$ e.u. Both $\Delta\overline{V}^*$ and $\Delta\overline{V}$ vary with temperature; $\Delta\overline{V}^* = 546.4 - 1.813T$ and $\Delta\overline{V} = -922.8 + 3.206T$. Because $ce^{\Delta S_o^*/R}$ is an inseperable constant, ΔS_o^* can not be evaluated.

The source of the increase in entropy for reversible protein denaturation is uncertain. It is possible that denatured proteins are less highly hydrated than native proteins and that release of "bound" water contributes to the

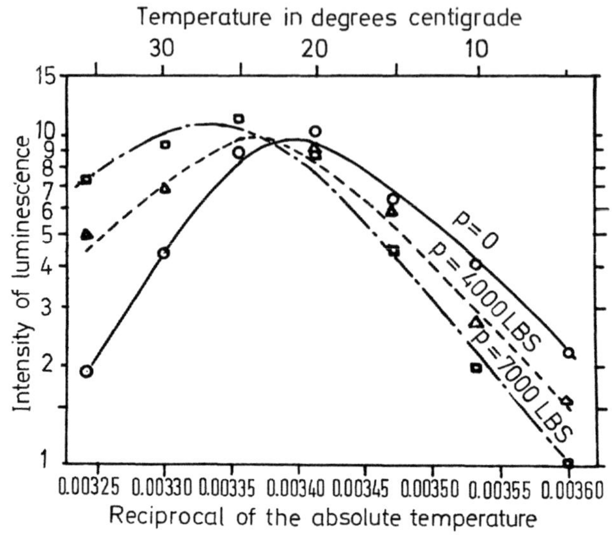

Fig. 2. The variation of luminescence intensity of *Photobacterium phosphoreum* with pressure and temperature. Ordinates are log luminescence intensity in arbitrary units. The curves are calculated from Eq. (10) with the parameters described in the text, and the points are the experimental data of BROWN *et al.* (33). From EYRING and MAGEE (34). Reprinted by permission of the copyright owner, The Wistar Institute Press

entropy increase. A probable source of entropy increase is change in the internal structure of the protein. Native protein molecules have a high degree of internal order with major regions of the polypeptide chain in the form of alpha helices. The polypeptide chain unfolds during the denaturation process. This transition from order to disorder contributes an entropy increase. In thermal denaturation experiments, the normal reaction product is a precipitate. It has long been realized that two separate reactions must be distinguished: conformational change or unfolding and coagulation of the unfolded product. The word denaturation is sometimes applied to the first step only and sometimes to the total process.

The first successful attempt to measure the volume changes in protein denaturation was that of HEYMANN (35). He found volume increases of 0.17 to 0.26 ml per 100 g for several proteins, including serum albumin, heated at their isoelectric points. However, it is necessary to distinguish between the changes accompanying unfolding or denaturation from those attributable to coagulation. HEYMANN found smaller positive volume changes at *pH* values removed from the isoelectric point, where coagulation does not occur. However, TANFORD (36) reviewed the literature on the effect of pressure on denaturation; small negative volume changes are indicated for ribonuclease in contrast with the parameter evaluation of EYRING and MAGEE for luciferase. JAENICKE (37) measured the volume changes in the heat aggregation of serum album at the isoelectric point. He demonstrated that no appreciable unfolding occurred. Bovine serum albumin in *pH*=4.7 acetate buffer at ionic strengths between 0.1 and 3.0 aggregates, as demonstrated by light scatter, when the temperature is raised to 65°C and reverts to the initial state when the temperature is reduced to 20°C. This aggregation is accompanyied by an increase in volume of approximately 0.12 ml per 100 g, or about 80 ml/mole. The volume increase was measured both by means of a dilatometer and in spring balance experiments similar to those performed on polymerizing TMV protein, as described in Chapter 2. Furthermore, hydrostatic pressure of 1250 atmospheres increases the temperature at which heat aggregation begins from about 56°C at atmospheric pressure to about 61°C at the higher pressure. High pressure also lowers the increment in aggregation per unit of temperature increase. Thus, in the respects investigated, isoelectric aggregation of bovine serum albumin resembles polymerization of TMV protein and the other proteins described in previous chapters. This is an entropy-driven reaction. The increase in volume and the increase in entropy upon aggregation can be explained, as in the other cases discussed, by the release of water molecules.

Interaction of DNA with Polyvalent Cations

ROSS and SHAPIRO (38) made calorimetric measurements of the interaction between DNA and the polyvalent cations, poly-(L-lysine), spermine and Mg^{++}. The experimental conditions for the reaction with the lysine polymer and spermine were such that the interaction with DNA is essentially complete. In the binding of Mg^{++}, three out of eight phosphate groups of the DNA were reacted. The calorimetrically determined ΔH expressed as cal/mole of phosphorus was -300, 0 and +350, respectively, for the binding of poly-(L-lysine), spermine and Mg^{++} to DNA. Since the interaction between DNA and these polyvalent cations occurs spontaneously, the very small negative or positive ΔH values obtained indicate that the binding of such cations is primarily an entropy-driven process.

Summary

This chapter presents a motley collection of entropy-driven, structure-forming processes for which relatively little detailed information is available. The one feature that these various processes have in common is that either all or much of the negative free energy of polymerization or structure formation comes from an increase in entropy. The purpose in presenting this material is to demonstrate the widespread and varied nature of entropy-driven processes. No claim is made that the list is exhaustive.

Most of the processes discussed involve the formation of structures by polymerization brought about by external interactions of proteins or other materials. In general, these reactions can be understood in terms of the principles discussed in Chap. 1. A notable exception is the helix-coil transition of poly-γ-benzyl-L-glutamate in ethylene dichloride-dichloro acetic acid mixtures. This is purely an internal rearrangement. The probable explanation is that, at low temperatures, interaction between solvent and the polymer chain is thermodynamically favorable, resulting in a random coil. At higher temperatures, this interaction is unfavorable; the chain assumes the α-helix configuration to avoid interaction with the solvent.

The binding of ions by proteins shares with polymerization reactions the feature that an increase in entropy accompanies an apparant decrease in the number of molecular species. The explanation for the entropy increase is that when ions are bound by the protein, the charges on both the ions themselves and the groups on the proteins to which they bind are neutralized. Water held by electrostriction is thereby released, providing the increase in entropy. Ion binding is, thus, an entropy-driven process which is not related to hydrophobic interactions.

Chapter 10

Entropy-Driven Processes in Biology

Introduction

Division of fertilized eggs from many species and formation of the spindle apparatus during cell division, protoplasmic streaming, pseudopodia formation in amoebae and color change in killifish are processes which either directly involve or are dependent upon entropy-driven structure-forming reactions. These processes have been reviewed briefly by LAUFFER (1), and more extensively by INOUÉ (2) and by MARSLAND (3).

Major contributions to our knowledge in this field have been made by MARSLAND and his associates. Two special pieces of apparatus were required for their research. One is a microscope-pressure chamber, first designed by MARSLAND and BROWN (4) and improved by MARSLAND (5). This consists of a pressure chamber with 400 ml capacity fitted with tempered plate glass windows, 7-10 ml thick, capable of withstanding pressure up to 20,000 pounds per square inch (psi). The pressure cell is mounted on the stage on an inverted microscope equipped with a long range objective. Magnifications up to 600 X are obtainable. With this assembly it is possible to observe living material while it is subjected to high pressures.

The other special apparatus is a pressure-centrifuge developed by BROWN (6). This consists of a T-shaped centrifuge rotor. In one arm of the rotor is a hollow space equipped with a needle valve. Biological material is placed in the hollow space, then the stem of the rotor is connected to a pressure apparatus and the desired pressure is applied. It is maintained by the needle valve. The other arm of the T contains a hollow space for control material at atmospheric pressure. The rotor is then screwed at the stem of the T onto a centrifuge. This apparatus permits simultaneous centrifugation of cells at atmospheric pressure and at higher pressures.

Both the microscope-pressure apparatus and the pressure-centrifuge were housed in a temperature control chamber described by MARSLAND (7). The temperature chamber is equipped with cooling and heating units, a fan and a thermo-regulator. Temperatures can be set at desired values between -5° and $60^{o}C$ with an estimated maximum error of $\pm 0.3^{o}C$.

These two pieces of equipment permit the investigator to observe directly the effects of temperature and pressure upon the behavior of living material and then to study the same material at the same temperature and pressure with the pressure-centrifuge. MARSLAND and his associates used the pressure-centrifuge to study the strength of the protoplasmic gel. The basis for this is that granules within the protoplasm fail to be displaced by centrifugation when the protoplasm is in a highly structured gel state. However, when the protoplasm is in an unstructured or sol state, the granules are displaced upon centrifugation. For example, the eggs of *Arbacia lixula* contain red chromato-

Fig. 1. Drawings of two eggs (*Arbacia lixula*) centrifuged simultaneously for 260 seconds at 34,000 X G and $15^{o}C$. The control egg (left), centrifuged at atmospheric pressure, displays many undisplaced pigment granules in the cortex of the hyaline zone. The experimental egg (right), which was centrifuged at 6000 $lbs./in.^{2}$ pressure, shows only a small number of undisplaced granules in the cortex, and represents the standard displacement end-point used in the experiments. From MARSLAND and LANDAU (8). Reprinted by permission of the copyright owner, The Wistar Institute Press

phores, a large proportion of which are localized in the cortex under the cell membrane. Most of these chromatophores remain stable during the early stages of cell division; when such cells are centrifuged, they are displaced. The time of centrifugation at constant speed required to remove most but not all of the granules from the cortex can then be taken as a measure of gel strength. The ratio of the time of centrifugation required by control cells at atmospheric pressure to that required by the cells under higher imposed pressures is a measure of relative gel strength. The extent of displacement is observed by stopping the centrifuge, chilling the material by dropping the pressure-centrifuge head into cold water and then releasing the pressure. The material is kept cold until observed with the microscope. The end point is illustrated in Fig. 1.

Fertilized Eggs

MARSLAND (7) studied the effects of temperature and pressure on the division of *Arbacia punctulata* eggs, using the microscope chamber. This was then correlated with the gel strength of the cortical protoplasm as measured with the pressure-centrifuge. The results are shown in Fig. 2. Cell division is inhibited by high pressure and low temperature and is favored by low pressure and high temperature. It is correlated with relative gel strength. The relative gel strengths in Fig. 2 are plotted on a logarithmic scale. The plots are reminiscent of the theoretical plots shown in Fig. 2 of Chap. 1. Thus, cleavage seems to be correlated with protoplasmic gelation, which is an endothermic process. From the effect of pressure shown in Fig. 2, one can calculate with the aid of Eqs. (4) and (5) of Chap. 9 that this gelation is accompanied by an increase in volume of approximately 100 milliliters per mole of protoplasmic material. If the molecular weight of the polymerizing substance is assumed to be 60,000, this volume increase corresponds to .0017 ml/g, a value considerably smaller than that for rod-forming polymerization of TMVP. MARSLAND and LANDAU (8) carried out similar experiments on eggs of four other species, *Arbacia lixula*, *Echinarachnius parma*, *Chaetopterus pergamentaceous* and *Rana pipens*, with similar results.

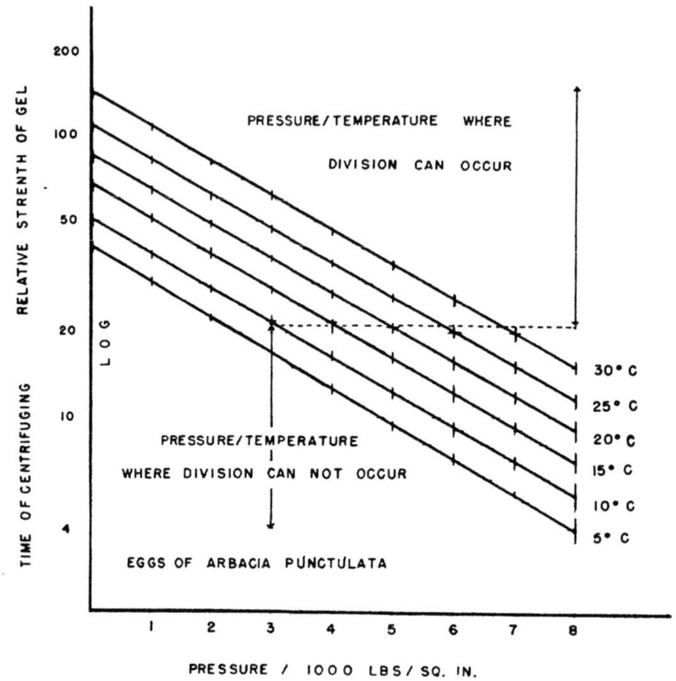

Fig. 2. Family of curves showing the effects of pressure, at different temperatures, on the structural strength of the gelated cortical protoplasm of the Arbacia egg, and the relation of these data to the capacity of the egg to perform the work of cleavage. Any combination of temperature and pressure which reduces the cortical gel strength beyond a critical value (relatively expressed as about 20 per cent of the atmospheric value at 25°C) is just adequate to block division. The centrifuge times are expressed directly in seconds, and the degree of variation is indicated by the length of the markers. From MARSLAND (7). Reprinted by permission of the copyright owner, The Wistar Institute Press

Formation of Pseudopodia in Amoebae

Another biological process which depends on an endothermic mechanism is the formation of pseudopodia by Amoebae (9). The effect at 25°C on *Amoeba dubia* is shown in Fig. 3. At high pressure, the Amoeba assumes an essentially spherical shape. When pressure is reduced, it recovers the ability to form

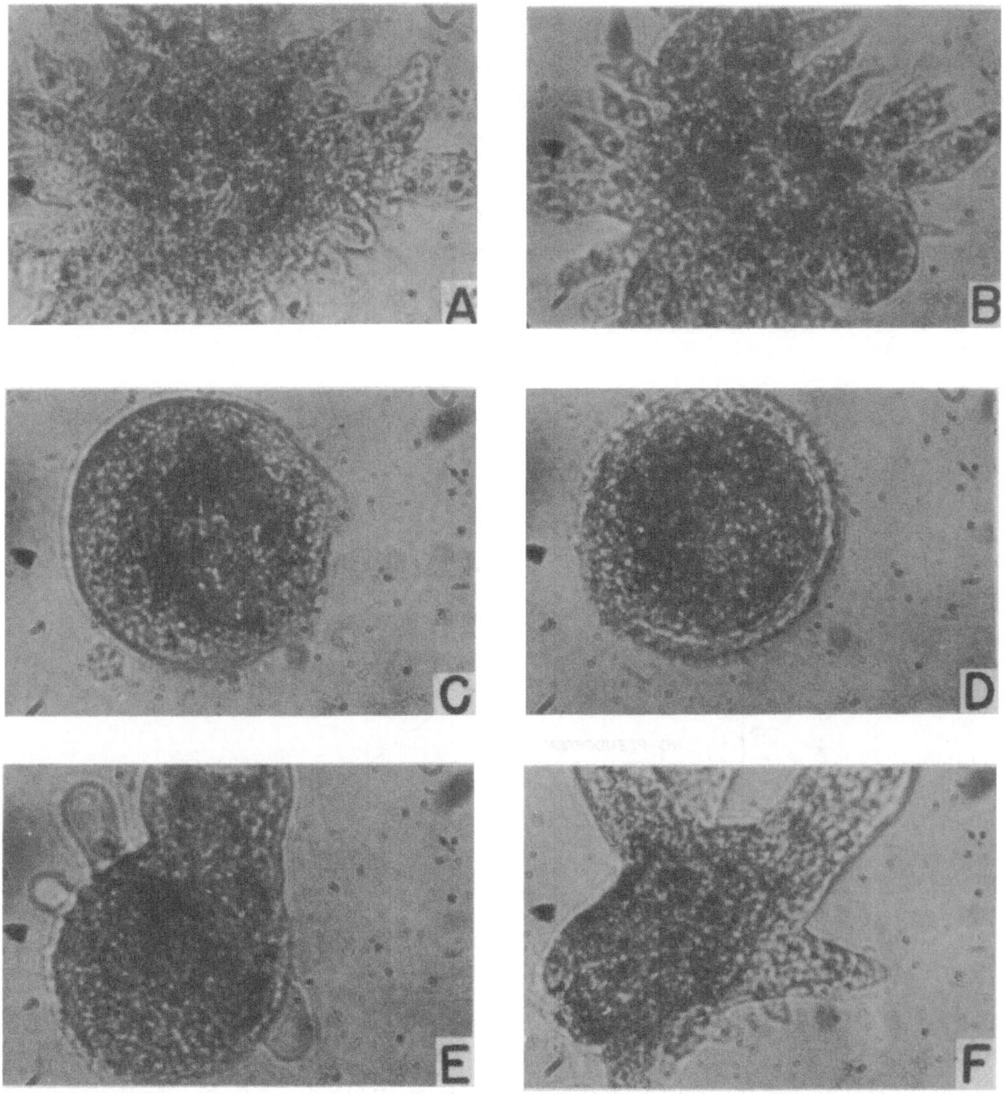

Fig. 3 A-F. Changes of form and activity in *Amoeba dubia* during and after exposure to a pressure of 6000 lbs./in.^{2} at 25°C. A. Normal form at atmospheric pressure. B. Ten seconds after pressure was raised to 6000 lbs./in.^{2}; pseudopodia beginning to retract. C. Pressure maintained for 5 minutes; rounding of specimen almost complete. D. Ten seconds after decompression. Note contraction of granular cytoplasm away from the cell membrane. E. Fifteen seconds later. Amoeboid activity recommencing. F. Five and a half minutes later. Amoeboid activity now quite vigorous. From LANDAU *et al.* (9). Reprinted by permission of the copyright owner, The Wistar Institute Press

pseudopodia. The effects of both temperature and pressure on *Amoeba proteus* are summarized in Fig. 4. This figure shows that the ability to form pseudopodia is also correlated with protoplasmic gel strength, which depends on temperature and pressure. At constant pressure, gel strength is higher at high temperature than at low temperature, and at constant temperature gel strength is increased by decreasing pressure. Comparable mechanisms must be involved in protoplasmic streaming investigated in the leaf cells of *Elodea canadensis* (10).

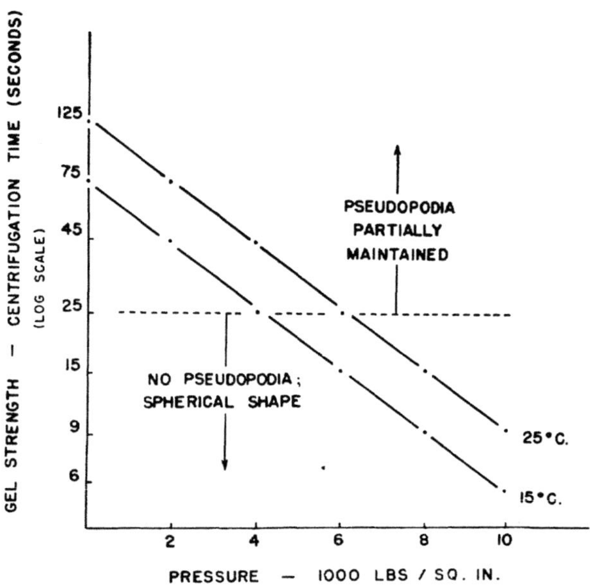

Fig. 4. Structural strength of plasmagel of *Amoeba proteus* plotted as a function of pressure at 15° and 25°C. Note that the two curves are parallel and that the critical gel strength, below which no pseudopodia can be maintained, is the same at both temperatures. Each point represents the average of values obtained in at least 6 determinations. From LANDAU *et al.* (9). Reprinted by permission of the copyright owner, Wistar Institute Press

Effects of ATP on Cleavage Capacity

From data such as those displayed in Fig. 2, it is possible to draw a graph showing the pressure level just sufficient to block cleavage at each temperature. LANDAU *et al.* (11) used this method to interpret the effect of ATP on furrowing strength. They added the high energy phosphate at a concentration of 0.0005 M to sea water containing the eggs, starting approximately 25 minutes before the onset of first cleavage. The results obtained with *Arbacia* are shown in Fig. 5. A similar strengthening of the gel by ATP was observed

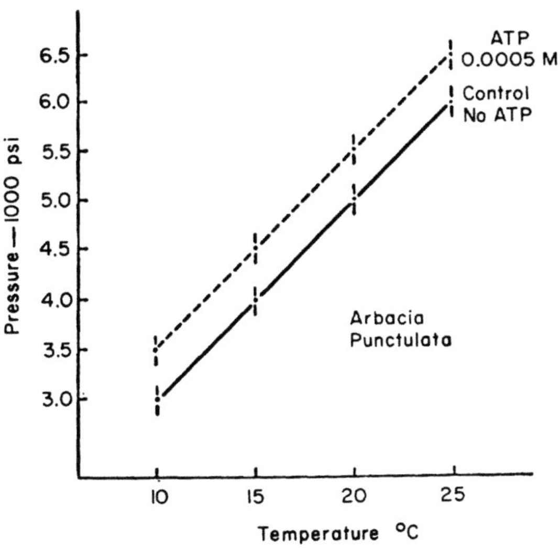

Fig. 5

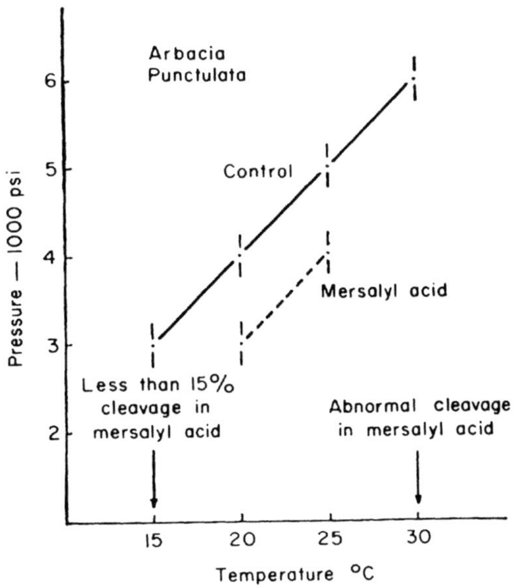

Fig. 6

Fig. 5. ATP-induced fortification of cleavage capacity. Each point indicates the pressure level which just suffices to block cleavage at each temperature. From MARSLAND (3) (Data originally from LANDAU *et al.* (11)). Reprinted by permission of the copyright owner, Academic Press, Inc.

Fig. 6. Reduction of cleavage potency by mersalyl acid (an inhibitor of ATP hydrolysis). Each point indicates the minimum pressure required to block the cleavage furrows. From ZIMMERMAN *et al.* (12). Reprinted by permission of the copyright owner, The Wistar Institute Press

for *Chaetopterus*. Mersalyl acid inhibits the hydrolytic splitting of the high energy phosphate bond of ATP. When this acid was added to produce a concentration of 0.004 M to the sea water containing the fertilized eggs 10 minutes after fertilization, the results shown in Fig. 6 were obtained (12). These two sets of experiments serve to indicate that hydrolysis of ATP contributes energy for building the gel structure of the cortical gel system.

MARSLAND (3) reported evidence from his own laboratory that substitution of H_2O by D_2O in the suspending medium of a cell enhances the gelation within the cell. Such studies were carried out on the cortical gel of cleaving eggs of *Arbacia punctulata*, on the plasma gel system of *Amoeba proteus* and on pigment movement and fine structure in the protoplasmic processes of fish melanocytes.

Color Changes in the Killifish

The melanophore in the skin of the killifish, *Fundulus heteroclitus*, is a star-shaped cell with arms or processes radiating out from the central region

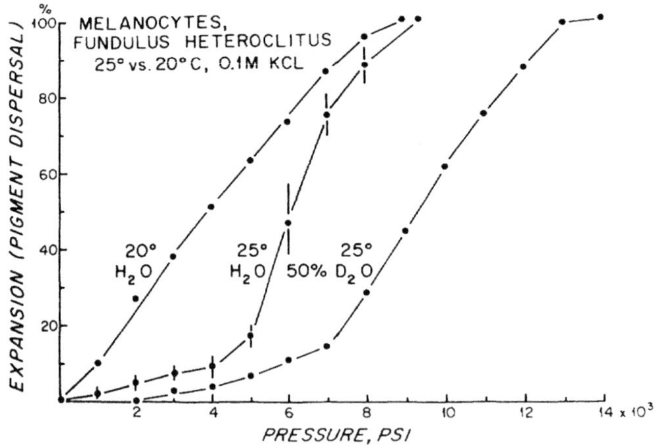

Fig. 7. Effects of temperature and D_2O on the pressure-induced expansion of melanocytes. Standard deviation values are shown by the vertical lines in the 25°C-H_2O curve. These values were maximal in the sense that the corresponding values of the other points, at each pressure level, were never greater. From MARSLAND and MEISNER (14). Reprinted by permission of the copyright owner, The Wistar Institution Press

containing the nucleus. When the melanin granules are in the central region of the cell, the fish appears pale, but when they are dispersed throughout the processes, they give the fish a dark color. Fig. 7 shows quantitatively the effects of temperature, pressure, and D_2O on pigment dispersal (14). High pressures and low temperatures favor dispersal of the pigment; thus dispersal seems to occur at low gel strength (13).

Karyokinesis

An early indication that the formation of the spindle-aster complex might be an entropy-driven reaction came from the work of PEASE. In 1941 he reported on the effects of hydrostatic pressure upon the spindle figure and chromosome movements in the first mitotic division of Urechis eggs (15). In 1946 he reported similar studies on the meiotic divisions of Tradescantia pollen mother cells (16). In these studies, PEASE found that movement of daughter chromosomes toward the poles could be retarded and finally stopped by applying pressures in the range, 4000-7000 psi. He also found that this effect was correlated with loss of structure in the spindle and astral regions. This he observed by fixing and sectioning the cells immediately after the release of the high pressure.

Further evidence, of a more direct nature, was obtained by INOUÉ on cells of *Chaetopterus pergamentaceous* and *Lilium longiflorum* undergoing mitotic division (17, 18). INOUÉ made direct observations using polarized light. This procedure permits spindles to be seen directly, as is shown in Fig. 8. These spindles disappear at temperatures below 10°C and are most strongly evident, as detected by birefringence, at temperatures of around 35°C. Therefore, spindle formation is an endothermic process.

INOUÉ made quantitative measurements of birefringence at different temperatures, shown in Fig. 9 as equilibrium retardation (Γ) as a function of temperature. He then interpreted his results on the assumption that there is an equilibrium between unoriented particles which do not exhibit birefringence and oriented particles of the same molecular weight which do exhibit birefringence. The equilibrium constant is given by Eq. (1),

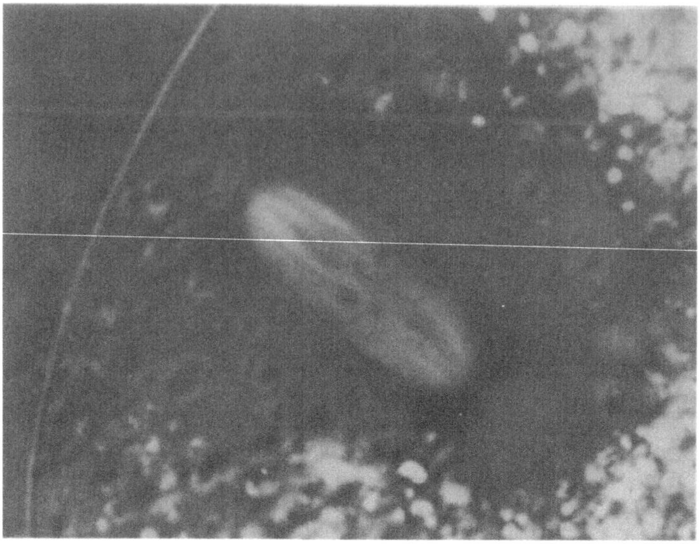

Fig. 8. Birefringent spindle fibers at metaphase in living cells of *Chaetopterus pergamentaceous*. From INOUÉ (18). Reprinted by permission of the copyright owner, The American Institute of Physics

$$K = B/(A_o - B) \tag{1}$$

where B is the transient concentration of oriented material, a function of temperature, and A_o is the maximum amount of oriented material obtained at high temperature. It is also the sum of the concentrations of oriented and unoriented material at any temperature. $(A_o - B)$ is therefore the transient concentration of unoriented material. INOUÉ's assumption that the equilibrium birefringence is directly proportional to B can be expressed by setting B equal to $\beta\Gamma$ and A_o equal to $\beta\Gamma_{max}$. The equilibrium equation then takes the form of Eq. (2).

$$K = \Gamma/(\Gamma_{max} - \Gamma) \tag{2}$$

INOUÉ assumed Γ_{max} is 6.0. When ln K thus determined is plotted against $1/T$ in accordance with Eq. (3) of Chap. 1, a straight line is obtained, as shown in Fig. 10. INOUÉ evaluated ΔH^o from the slope to be 28.4 kcal/mole and calculated ΔS^o to be +101 entropy units.

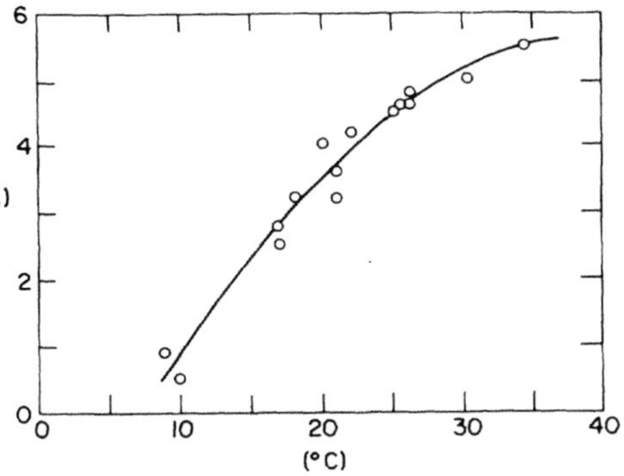

Fig. 9

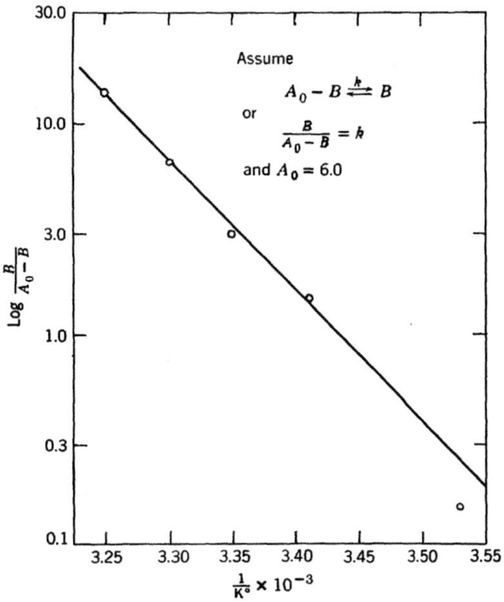

Fig. 10

Fig. 9. Equilibrium retardation (Γ) of Chaetopterus spindle at various temperatures. From INOUÉ (18). Reprinted by permission of the copyright owner, The American Institute of Physics

Fig. 10. Log plot of spindle reaction equilibrium versus inverse absolute temperature. From INOUÉ (18). Reprinted by permission of the copyright owner, The American Institute of Physics

The model used by INOUÉ to interpret his results seemed to be reasonable enough at the time that his work was done. However, as will be shown later in this chapter, the unoriented material in the pool from which the spindles are formed is the protein, tubulin, and in the unoriented condition it is unpolymerized. Therefore, a biologically much more plausible assumption is that n molecules of unoriented unpolymerized protein, U, are in equlibrium with oriented polymerized material, P. The equilibrium is now expressed by Eq. (3).

$$K = [P]/[U]^n \tag{3}$$

By making INOUÉ's assumption concerning the relationship between concentration and birefringence, one can set [P] = βΓ, which is in turn equal to

$\beta\Gamma_{max}\alpha$, where α varies with temperature from approximately 0.1 to 1. The mass action equation now becomes Eq. (4). The approximation on the right is very accurate when n is greater than about 100.

$$K^{1/n} = (\beta\Gamma_{max})^{1/n}\, a^{1/n}/[\mathrm{U}] \simeq (\beta\Gamma_{max})^{1/n}(1/[\mathrm{U}]) \tag{4}$$

Eq. (3) of Chap. 1 can be written in the form of Eq. (5).

$$(1/n)\ \ln K = (1/n)(\Delta S^{o}/R) - (1/n)(\Delta H^{o}/R)(1/T) = (\Delta S^{o'}/R) - (\Delta H^{o'}/R)(1/T) \tag{5}$$

In this equation, ΔS^{o} and ΔH^{o} are the entropy and enthalpy per mole of polymerized material formed, and $\Delta S^{o'}$ and $\Delta H^{o'}$ are the corresponding parameters per mole of unoriented unpolymerized pool material used. When Eq. (4) is substituted into Eq. (5), Eq. (6) results.

$$-\ln[\mathrm{U}] = (\Delta S^{o'}/R) - (1/n)\ \ln(\beta\Gamma_{max}) - (\Delta H^{o'}/R)(1/T) \tag{6}$$

Except for differences in notation and the second term on the right, which approaches zero as n becomes very large, this equation is identical with Eq. (2) of Chap. 8, the equation for a phase transition. In accordance with the assumption that concentration is directly related to birefringence, [U] can be set equal $n\beta(\Gamma_{max} - \Gamma)$. When this is substituted into Eq. (6), Eq. (7) results.

$$-\ln(\Gamma_{max} - \Gamma) = \ln(n\beta) - (1/n)\ \ln\ (\beta\Gamma_{max}) + (\Delta S^{o'}/R) - (\Delta H^{o'}/R)(1/T) \tag{7}$$

Since the first three terms on the right of Eq. (7) are all constants, one should obtain a straight line when $-\ln\ (\Gamma_{max} - \Gamma)$ is plotted against $1/T$, and from the slope one can evaluate $\Delta H^{o'}$. INOUÉ's data shown in Fig. 9 are thus plotted in Fig. 11. For the purpose of this plot it was assumed that the data of Fig. 9 extrapolate to a value of Γ_{max} of 6.3, an assumption which is neither more nor less reasonable than INOUÉ's original assumption that the extrapolated value is 6.0. It is clear that a straight line fits within the reproducibility of the data. From the slope of this straight line a value of 14.3 kilocalories per mole of unpolymerized material can be calculated for $\Delta H^{o'}$. The intercept of the graph can readily be calculated to be +23.8. This constant, however, is not equal to $\Delta S^{o'}/R$ because, as is shown in Eq. (7), it is the sum of three terms. However, the second term on the right of Eq. (7)

approaches zero when n becomes very large. Furthermore, the concentration of U before any polymerization takes place, $[U]_o$, is equal to n $\beta\Gamma_{max}$. Since the molar concentration of unpolymerized protein must be a very small fraction of 1 mole per liter and since Γ_{max} is 6.3, $n\beta$ must be a small fraction and its logarithm must be a negative number. Therefore, one can say at least that $\Delta S^{o'}$ must be positive and greater than 23.8 x 2 or 47.6 entropy units. Thus, it is clear that the formation of spindles is an entropy-driven process.

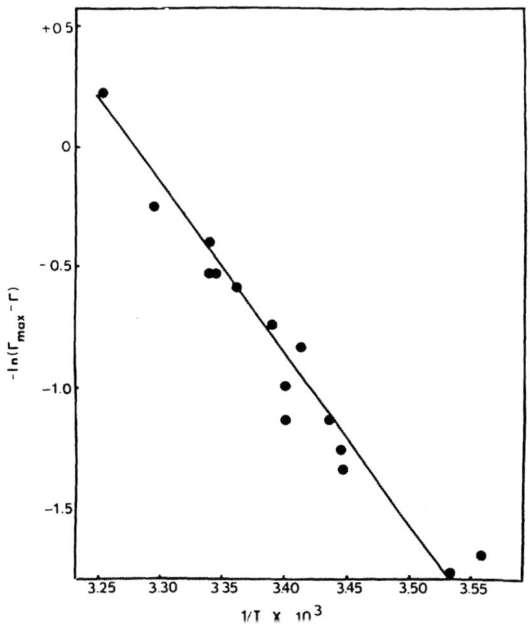

Fig. 11. Interpretation of the data of Fig. 9 in accordance with Eq. (7) on the assumption that Γ_{max} has a value of 6.3

ZIMMERMAN and MARSLAND (19) exposed the eggs of *Arbacia punctulata* to pressures in the range 2000-10,000 psi, applied at various stages in first cleavage. Immediately following, they isolated spindle-aster-chromosome complexes. Their isolation procedure permits the separation of intact spindle-aster-chromosome complexes from dividing cells not treated with pressure. However, pressure treatment resulted in a drastic disorganization of the linear and radial structures in the extracted material. If the pressure was released and sufficient time allowed, normal structures could be isolated.

Effect of Heavy Water on the Mitotic Apparatus

INOUÉ *et al.* (20) found that when developing eggs of the sea urchin, *Lytechinus variegatus*, are placed in sea water containing 30-40% D_2O, the birefringence of the spindle fibers rose above normal, but mitosis proceeded. At higher concentrations of D_2O, the spindle birefringence rose even higher, but mitosis was arrested. The increase in birefringence, showing increased polymerization, parallels the effect of heavy water on the polymerization of tobacco mosaic virus protein, described in Chap. 4. MARSLAND and ZIMMERMAN (21) isolated mitotic apparatuses, using the method of MAZIA and DAN (22), following treatment with both D_2O and pressure. Eggs immersed in 90% D_2O and then subjected to pressures of 12,000 psi for three minutes yielded isolated mitotic structures not distinguishable from those obtained from control cells kept in water and not pressurized. In contrast, pressurized nondeuterated eggs have isolates void of visible structure. Even with 50% D_2O, substantial stabilization was observed.

In spite of the fact that D_2O strengthens the mitotic apparatus, it blocks mitosis when present in the cell at levels of 70% or greater. MARSLAND (3) suggested that "the most probable answer seems to be that such deuteration over-stabilizes the gel structure of the mitotic apparatus, preventing the ebb and flow of polymerizations that seem to be essential to the continuance of mitotic activity." Since high pressure has an opposite effect, MARSLAND reasoned that pressure might be able to oppose not only the structural effects of D_2O upon the mitotic apparatus, but also its inhibitory effects upon mitotic activity. MARSLAND reviewed a number of experiments from his own and from INOUÉ's laboratory which support this reasoning. For example, in 80-90% deuterated sea water, no eggs of *Stronglocentrotus purpuratus* showed any attempt to cleave when immersed during early prophase. At pressures between 2500 and 5000 psi 98% they developed furrows and many completed first cleavage.

Effects of Colchicine

The well-known antimitotic action of colchicine probably results from a weakening of the structure of the mitotic spindle. INOUÉ (2) found that colchicine lessens the birefringence in the spindle of dividing eggs. MARSLAND (23) studied the pressures required to inhibit first cleavage of the eggs of *Lytechinus variegatus*. His results are shown in Figure 12. It is clear that pressure and colchicine act synergistically, that is, both inhibit spindle formation. The effect of colchicine is antagonized by 25-35% concentrations of D_2O, even though the opposite effect was found at higher concentrations (24). INOUÉ *et al.*, (25) found that D_2O and Colcemid have opposite effects upon the birefringence of the mitotic spindle. Preliminary studies carried out in the author's laboratory indicate that colchicine had little or no effect on the polymerization of TMVP (26).

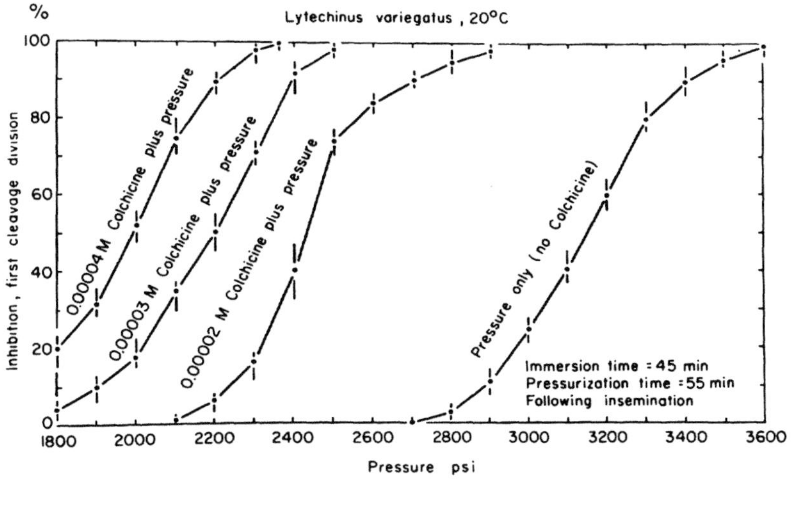

Fig. 12. Inhibition of first cleavage by colchicine and pressure applied concurrently. Each point represents the average of at least three experiments in which about 300 eggs were observed. Vertical lines indicate extremes of variation. From MARSLAND (23). Reprinted by permission of the copyright owner, The Wistar Institute Press

The Axopodia of Heliozoa

TILNEY *et al.*, (27) studied the effect of hydrostatic pressure on the axopodia of the heliozoan, *Actinosphaerium nucleofilum* . Axopodia are needle-like structures which radiate out from the cell surface of *Actinosphaerium*. This

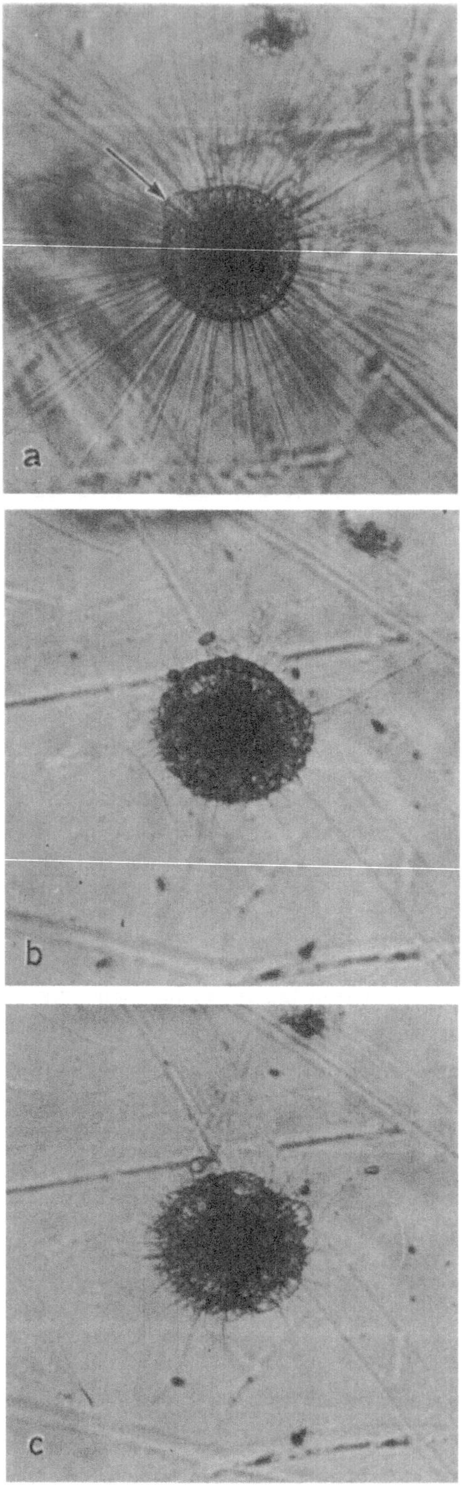
a
b
c

organism was studied in MARSLAND's pressure chamber. As shown in Fig. 13a, before pressure is applied, the needle-like Axopodia extend from the cell body into the culture medium. Fig. 13b shows that after 10 minutes of pressure at 4,000 psi, only short stubs of the axopodia remain. After 20 minutes of sustained pressure at 4,000 psi, the axopodia become stabilized at a somewhat greater length than that shown after 10 minutes. This is illustrated in Fig. 13c.

Relationship of Endothermic Cellular Processes to Microtubules

The pressure-temperature effects exhibited in the various cellular processes herein described are strongly correlated with polymerization-depolymerization of tubular protein into microtubules, a process which resembles the polymerization-depolymerization of TMVP. Knowledge about microtubules and their relationship to the mitotic apparatus and the cortical gel developed gradually. HARRIS observed microtubules in the mitotic apparatus of sea urchin embryos (28). LEDBETTER and PORTER fixed root tips of *Phleum pratense L* seedlings, whole plants of *Spirodela oligorrhiza Kurtz* and root tips from cuttings of *Juniperus chinensis L* for one and a half hours in 6% gluteraldehyde and then examined the material with the electron microscope (29). Slender tubules, 230-270 Å in diameter, were found in the cortical regions and in mitotic

---

◀ Fig. 13 a-c. Light micrographs taken of the same living specimen of *Actinosphaerium nucleofilum* at a magnification of X 100. (a) Specimen in the pressure chamber prior to compression. Needlelike processes, or axopodia, extend from the cell body or cytosome into the culture medium. The cytosome is divided into an outer cortical region, in which a single contractile vacuole is present (see arrow), and an inner denser medullary region. The axopodial axis, or axoneme, continues into the latter. (b) Micrograph taken after 10 min of sustained pressure at 4,000 psi. Only short stubs of the axopodia remain. (c) Micrograph taken after 20 min of continuous pressure at 4,000 psi. The axopodia have become twice as long as those shown in the previous photograph. From TILNEY *et al.* (27). Reprinted by permission of the copyright owner, the Rockefeller University Press

spindles. DE THÉ found microtubules, 180-250 Å in diameter and of undetermined length, in avian, murine and human cells fixed with gluteraldehyde (30). These cytoplasmic microtubules in animal cells were connected with the satellites of the centrioles and were reported to be similar to mitotic spindle fibers. Further evidence for microtubules in the mitotic apparatus was found for dividing cells of the European corn borer (*Ostrinia nubilalis*), the roach (*Periplaneta americana*), *Tradescantia paludosa* stamens, root tips of *Raphanus sp.* and the giant amoebae (*Pelomyxa carolinensis*) by ROTH *et al.* (31).

The relationship between the formation and the disappearance of microtubules and the formation and disappearance of the mitotic apparatus was investigated by KANE and FORER (32). The mitotic apparatus was isolated and then stored at various temperatures. Polarization microscopy showed a decrease in birefringence upon storage, and electron microscopy showed that the tubular filaments at 200 Å in diameter which formed the spindle of the intracellular mitotic apparatus broke down during storage to form rows of dense granules. INOUÉ and SATO reported studies on the spindle of *Pectinaria oocytes* in which a new fixative was used, a mixture of gluteraldehyde and hexylene glycol (33). With this fixative, the birefringence of fixed mitotic apparatuses was the same as before fixing. The elevated birefringence in cells treated with D_2O and the reduced birefringence of cells treated with Colcemid retained their values when fixed. Finally, the number of spindle filaments observed with the electron microscope showed a close correlation with the birefringence of the spindle fiber for D_2O treated eggs and their untreated controls. The structure of microtubules from the mitotic apparatus was discussed by KIEFER *et al.* (34). They visualized the tubule as consisting of a cylindrical alignment of fibrils about 35 Å in diameter with a center to center spacing of about 50 to 60 Å. The filaments are beaded, the beads in the parallel filaments being in register. In agreement with previous work by others, KIEFER *et al.*, found 13 fibrils in a microtubule.

Both KIEFER *et al.* and STEPHENS (35) have isolated and characterized the protein from the mitotic apparatus. Somewhat different results were reported, possibly related to differences in the method of isolation. Both agree that the smallest particle found has a sedimentation coefficient of about 2.5 S; this should have a diameter in the neighborhood of 40 Å, about right for the diameters of the beads in the fibrils. Both studies also reveal particles which sediment at about 22 S with various intermediates.

There are apparently many kinds of microtubule. STEPHENS reported that the proteins obtained from the microtubules of the outer fibers of flagella and cilia differ from that obtained from the microtubules of the mitotic apparatus in that they polymerized in a manner similar to that of bacterial flagella; nucleation is necessary. Furthermore, low temperatures did not appreciably inhibit polymerization, in contrast with the *in vivo* results on the polymerization of the protein from the microtubules of the mitotic apparatus (36). In a later report (37), STEPHENS described isolation of two somewhat different proteins from the A-subfiber and B-subfiber of the outer fiber double microtubules of cilia and flagella. TILNEY and GIBBONS (38) showed that at each step of development of the primary mesenchyme in sea urchin embryos, microtubules are disposed in patterns correlating with the shape of the cell at that stage. Colchicine and pressure break down the microtubules, and the cells resort to a spherical shape. Heavy water stabilizes the microtubules and the shape of the cells.

Further evidence for the relationship of microtubules to entropy-driven cellular processes comes from work with the killifish (*Fundulus heteroclitus L*) done by BIKLE *et al.* (39), who studied the pigment migration. The pigment migrates along rather fixed paths. The paths or processes remain, even when the granules are in the contracted form. The zones of cytoplasm around these channels are populated by microtubules about 225 Å units in diameter, aligned parallel to the direction of pigment movement. These long, slender elements are found whether the melanin is contracted in the cell center or dispersed. BIKLE *et al.* reasoned that the microtubules function as cytoskeletal elements which both define the channels in which the pigment moves and help to maintain the extended form of the melanophore arms. They postulated that microtubules exert a motive force on the pigment granules and suggest a model based on an interaction between the surfaces of the granules and the microtubules.

Perhaps the best demonstration of the relationship between microtubule formation and disintegration with entropy-driven processes in cells comes from the studies on the axopodia of heliozoa. TILNEY and PORTER (40) described the structure of the axopodia of *Actinosphaerium nucleofilum*. The axopodia are more than 400 μ long and 5-10 μ in diameter. Each has a central birefringent core, called an axoneme, which runs from the medulla to the tip. Each axoneme is made up of a bundle of parallel microtubules. These are 210-240 Å in diameter and are arranged in a double coil, as illustrated in Fig. 14. The separation between tubules in rows is 70 Å, and rows are separated by 300 Å. The number of tubules is greatest at the axopod base and smallest at the tip. The tubules at the center run the full distance, and those near the periphery at the axopod base run the shortest distance; the nearer the center of the axopod, the longer are the microtubules. When specimens of this organism are chilled to 4°C, the axopodia retract and the birefringent core disappears (41). After the organism is returned to room temperature, the axonemes reappear and the axopodia re-form. Microtubules organized the same way as in untreated specimens are always found in each growing axopod. Furthermore, as shown by the work of TILNEY *et al.* (27), the application of hydrostatic pressure results in drastic shortening of the axopodia. As illustrated in Fig. 15, this is also correlated with disintegration of the microtubules.

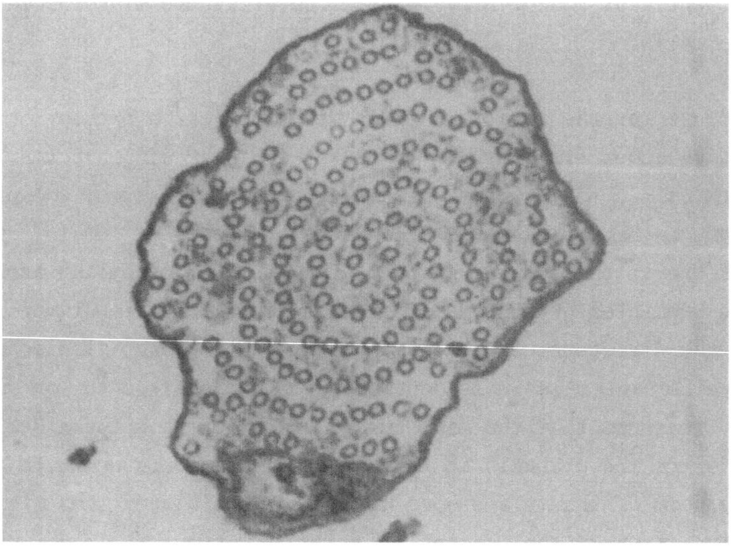

Fig. 14. Cross-section through an axopodium of an unpressurized control specimen. The microtubules, which form the axoneme, run parallel to each other and are arranged in an interlocking double coil Fixation, 1% glutaraldehyde. x75,000, Courtesy of Dr.L.G. TILNEY. Similar to figure published by TILNEY *et al.* (27).

In a review published in 1971, STEPHENS (42) summarized what was known about microtubules. Most of our knowledge was based on studies of ciliary, flagellar and neurotubular microtubules. Only sparse chemical information was available

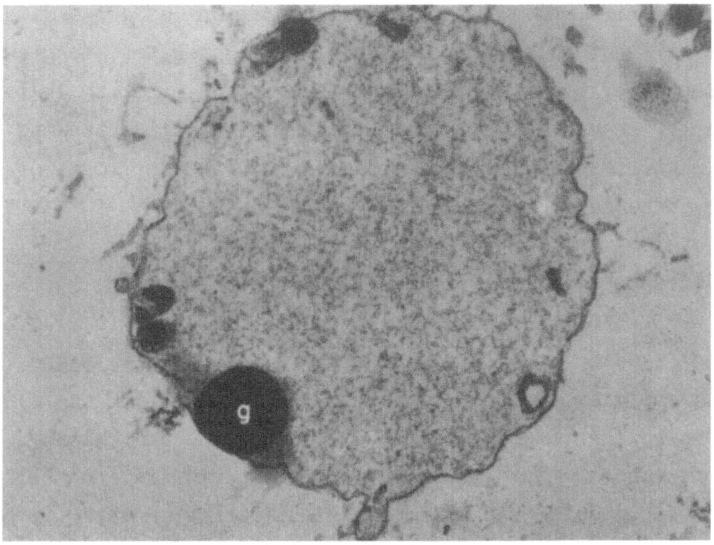

Fig. 15. Cross-section through one of the short axopodia present after ten min at 4000 psi. The central region normally occupied by the axoneme contains finely fibrillar material. Peripheral to this, beneath the unbroken plasma membrane, are several electron-opaque granules (*g*). Fixed with 1% glutaraldehyde. x35,000. From TILNEY *et al.* (27). Reprinted by permission of the copyright owner, The Rockefeller University Press

on cytoplasmic and mitotic microtubules and none on axopodial systems. From the available data, however, one can state that microtubules are composed of tubulin, a 40 Å globular protein with a molecular weight of 55-60,000. From X-ray diffraction and electron microscope studies, it is concluded that the subunits are arranged in a 40 by 53 Å surface lattice with 12-or 13-fold rotational symmetry. The microtubules in the outer fiber doublet of flagella yield two distinct tubulins which differ somewhat in amino acid composition.

In many ways tubulin resembles actin. However, tubulin and actin from the same species differ significantly from each other; there is greater resemblance to homogolus proteins from other species. Furthermore, the presently accepted molecular weight of actin is 47,000, while that for tubulin is about 60,000.

Tubulin seems to exist under physiological conditions as a 6 S dimer. The monomer alone will not polymerize unless higher aggregates are present to act as nucleii. The dimer binds colchicine and other antimitotic alkaloids. Colchicine evidently prevents tubule formation by competitive binding. In contrast, Vinca alkaloids lead to the formation of different forms of polymerized tubulin at the expense of microtubules.

While early attempts to polymerize tubulin *in vitro* by changing the temperature failed, WEISENBERG (43) found that tubulin isolated from rat brain can be polymerized *in vitro* in solutions containing ATP or GTP, magnesium ion and a good calcium chelator. Polymerization takes place at 35°C and can be reversed by dropping the temperature to 0°C. Colchicine and calcium ions inhibit polymerization; magnesium ion is essential for polymerization. Microtubule formation does not require the addition of nucleating centers. Since aggregates too small to be resolved by electron microscopy might be present in the low-molecular weight form, this result is not necessarily in contradiction to

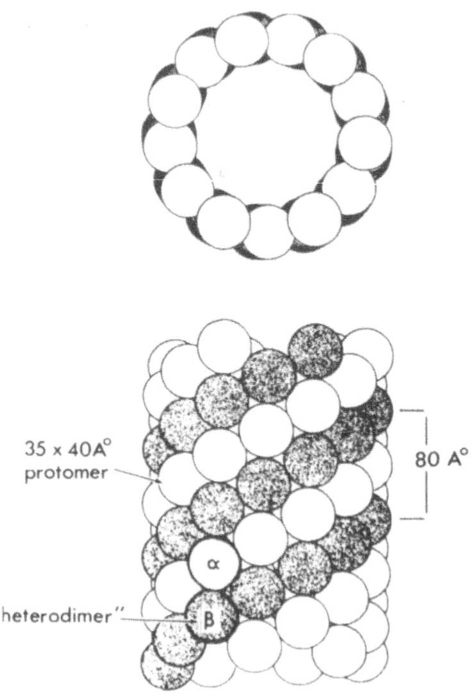

Fig. 16. Microtubule model. From BRYAN (44). Reprinted by permission of the copyright owner, Federation of American Societies for Experimental Biology

that reported above. The extreme sensitivity of tubulin to calcium ions, which inhibit, explains the failure of previous efforts to obtain polymerization. This, combined with the promoting action of magnesium ions, suggests a possible mechanism for regulation of microtubule polymerization *in vivo*.

Research on purification and properties of tubulin and its polymerization into microtubules has continued at an accelerated pace. Three short reviews (44, 45, 46) describe the status in mid 1973. The finding of WEISENBERG that tubulin undergoes entropy-driven polymerization *in vitro* was confirmed. Clear evidence is available that the tubulin dimer is composed of one molecule each of chemically different α-tubulin and β-tubulin. Partial amino acid sequences for both were reported. A proposed structure of the microtubule is reproduced as Fig. 16.

Summary and Conclusions

Data presented in this chapter demonstrate that important biological processes, such as division of fertilized eggs, formation of the spindle apparatus during cell division, pseudopodia formation in amoebae, color change in killifish and protoplasmic streaming are entropy-driven. Evidence that they are endothermic derives from the fact that they proceed at high temperatures and not at low temperatures. In common with the entropy-driven reactions previously discussed, these processes involve an increase in volume, in some instances measured directly and in other instances inferred from the inhibitory action of high pressure. Most of these processes have been correlated experimentally with strength of protoplasmic gel. The processes occur when the gel is strong and do not when it is weak.

Underlying each is a chemical reaction, the reversible polymerization of the protein, tubulin, to form rod-like microtubules. These microtubules cause protoplasm to change from the sol state to the gel state and they form the spindle fibers found in dividing cells. The polymerization of tubulin is an endothermic, entropy-driven reaction, resembling in many ways the polymerization of tobacco mosaic virus protein, TMVP. Since the increase in entropy needed to drive TMVP polymerization comes from the release of water molecules, it is appropriate to conclude that release of water molecules provides the entropy increase needed for the polymerization of tubulin. Water is thus seen to be more than a mere vehicle for the biochemical reactions necessary for life; it is a crucial contributor to biological structures and processes. In a sense, the main theme of this book is an emphasis on the newly appreciated role for water in living systems.

Biological processes of the sort discussed here involve dramatic changes in the structure of the cell, changes which must be turned on when a function, like cell division, is to be performed and turned off when that function is no longer needed. Entropy-driven structure-forming reactions which derive their entropy increase from release of water molecules are ideally suited for this requirement. In the first place there is plenty of water available in the cell to be "bound" and released. Actually, very few water molecules need to be released. In the best investigated example, the polymerization of TMVP, 0.027 g water are released per gram of protein. This calculates out to be only 26 water molecules per chemical repeat unit of protein, about 1/10 of the total "bound" water available. Furthermore, these reactions have the remarkable property that the equilibrium can be shifted from one direction to the other by almost trivial changes in the immediate environment. In 0.01 M phosphate buffer at room temperature, TMVP is almost totally unpolymerized at pH=6.35, and almost completely polymerized at pH=6.25. The difference amounts to a change in hydrogen ion activity in the immediate region of the protein molecule from 4.5×10^{-7} to 5.5×10^{-7} mole/l. This increase of 1×10^{-7} mole/l represents an addition of only one hydrogen ion in 1/60 μ^3 or, when the protein concentration is 1mg/ml, only one hydrogen ion in a volume containing over 500 protein subunits. This is indeed a small change in environment, easily brought about; it is a change from one in which protein is unpolymerized to another in which it is polymerized. This change in environment is, of course, in addition to the approximately one hydrogen ion per protein monomer bound during polymerization under these conditions.

The author's concern, both in the laboratory and in the library, for the role of water in biological systems originated when he was the student of R.A. GORTNER. GORTNER had an appreciation of such matters far in advance of his time and even of the data available to him. In 1930, when water was still considered to be primarily a solvent for biological molecules and a supporting medium for organism, GORTNER (47) wrote; "The biocolloids in general belong to the lyophilic group, and their actions and interactions are determined not alone by the chemical nature of the ultimate dry constituents but rather by their properties in the solvated state". This is the message of this book - the behavior in the solvated state of biologically important proteins: tobacco mosaic virus protein, sickle cell hemoglobin, collagen, myosin, actin and tubulin.

References

Chapter 1

1. LAUFFER, M.A.: Protein-Protein Interaction; Endothermic Polymerization and Biological Processes. In: Proteins and Their Reactions (H.W.SCHULTZ, A.F.ANGLEMIER, eds.), chap.5. Westport/Conn.: Avi Publ.Co.1964.
2. HAMBLY, A.N., BYRNE, R.T.: A Spontaneous Endothermic Reaction, J.Chem. Educ.46, A55 (1969).
3. SZEGVARI, A.: The Gelation of Nitrocellulose Solutions. Kolloid Z.34, 34-37 (1924).
4. HEYMAN, E.: Studies on Sol-Gel Transformation. I. The Inverse Sol-Gel Transformation of Methylcellulose in Water. Trans.Faraday Soc.31, 846-864 (1935).
5. FREUNDLICH, H.: Some Recent Work on Gels. J.Phys.Chem.41, 901-910 (1937).
6. STEVENS, C.L., LAUFFER, M.A.: Polymerization-Depolymerization of Tobacco Mosaic Virus Protein. IV. The Role of Water. Biochemistry 4, 31-37 (1965).
7. KAUZMANN, W.: Some Factors in the Interpretation of Protein Denaturation. Advan.Protein Chem.14, 1-63 (1959).
8. FRANK, H.S., EVANS, M.W.: Free Volume and Entropy in Condensed Systems. III. Entropy in Binary Liquid Mixtures; Partial Molal Entropy in Dilute Solutions; Structure and Thermodynamics in Aqueous Electrolytes. J.Chem. Phys.13, 507-532 (1945)
9. BEN-NAIM, A.: Thermodynamics of Dilute Aqueous Solutions of Nonpolar Solutes. In: Water and Aqueous Solutions: Structure, Thermodynamics, and Transport Processes (R.A.HORNE, ed.), pp.425-464. New York: Wiley-Interscience 1971.
10. TANFORD, C.: Physical Chemistry of Macromolecules. New York: Wiley 1961.
11. LAUFFER, M.A.: Polymerization-Depolymerization of Tobacco Mosaic Virus Protein.VII. A Model. Biochemistry 5, 2440-2446 (1966).
12. LUMRY, R., RAJENDER, S.: Enthalpy-Entropy Compensation Phenomena in Water Solutions of Proteins and Small Molecules: A Ubiquitous Property of Water. Biopolymers 9, 1125-1227 (1970).
13. STEVENS, C.L., FELSENFELD, G.: The Conversion of Two-Stranded Poly (A + U) to Three-Strand Poly (A + 2U) and Poly A by Heat. Biopolymers 2, 293-314 (1964).
14. EDSALL, J.T.: Dipolar Ions and Acid-Base Equilibria. In: Proteins Amino Acids and Peptides (E.J.COHN, J.T.EDSALL, eds.), chap.4, pp.75-115. New York: Reinhold 1943.
15. EDSALL, J.T.: Some Relations Between Acidity and Chemical Structure. In: Proteins Amino Acids and Peptides (E.J.COHN, J.T.EDSALL, eds.), chap.5, pp.116-139. New York: Reinhold 1943.
16. COHN, E.J., EDSALL, J.T.: Interactions Between Organic Solvents and Dipolar Ions Estimated from Solubility Ratios. In: Proteins Amino Acids and Peptides (E.J.COHN, T.J.EDSALL, eds.), chap.9, pp.196-216. New York: Reinhold 1943.

17. SIMMONS, N.S., BLOUT, E.R.: The Structure of Tobacco Mosaic Virus and Its Components: Ultraviolet Optical Rotatory Dispersion. Biophys.J., 1-55 (1960).

18. JAENICKE, R., LAUFFER, M.A.: Polymerization-Depolymerization of Tobacco Mosaic Virus Protein. XII. Further Studies on the Role of Water. Biochemistry 8, 3083-3092 (1969).

19. KHALIL, M.T.M.: Optical Rotatory Dispersion and Circular Dichroism Study of the Polymerization-Depolymerization of TMV Protein. (in preparation) (1972).

20. SCHUBERT, D., KRAFCZYK, B.: Circular Dichroism Studies on the Reversible Depolymerization of Tobacco Mosaic Virus Protein, Biochim.Biophys.Acta. 188, 155-157 (1969).

21. URRY, D.W., HINNERS, T.A., MASOTTI, L.: Calculation of Distorted Circular Dichroism Curves for Poly-L-Glutamic Acid Suspensions. Arch.Biochem. Biophys.137, 214-221 (1970).

22. JAENICKE, R.: Volume Changes in the Isoelectric Heat Aggregation of Serum Albumin. Eur.J.Biochem.21, 110-115 (1971).

23. BUDZYNSKI, A.Z.: Circular Dichroism Studies on Coat Protein of Some Strains and Mutants of Tobacco Mosaic Virus. Biochem.Biophys.Acta,251, 292-302 (1973).

24. JAENICKE, R., VOGEL, D.: Conformational Changes and Proton Uptake in the Reversible Aggregation of Tobacco Mosaic Virus Protein. Eur.J.Biochem.41, 607-615 (1974).

Chapter 2

1. STEVENS, C.L., LAUFFER, M.A.: Polymerization-Depolymerization of Tobacco Mosaic Virus Protein IV. The Role of Water. Biochemistry 4, 31-37 (1965).

2. JAENICKE R., LAUFFER, M.A.: Polymerization-Depolymerization of Tobacco Mosaic Virus Protein XII. Further Studies on the Role of Water. Biochemistry 8, 3083-3092 (1969).

3. LAUFFER, M.A.: Polymerization-Depolymerization of Tobacco Mosaic Virus Protein II. Theory of Protein Hydration. Biochemistry 3, 731-736 (1964).

4. ADAIR, G.S.: The Theory of Membrane Equilibrium. Trans.Faraday Soc.33, 1106-1116 (1937).

5. LAUFFER, M.A.: Protein-Protein Interaction: Endothermic Polymerization and Biological Processes. In: Symposium of Foods; Proteins and Their Reactions (H.W.SCHULTS, A.F.ANGLEMIER, eds.), Chap.5, pp.87-116. Westport/Conn.: Avi Publ.Co.1964.

6. FINCH, J.T.: The Hand of the Helix of Tobacco Mosaic Virus. J.Mol.Biol., 66, 291-294 (1972).

7. SCHEELE, R.B., LAUFFER, M.A.: Acid-Base Titrations of Tobacco Mosaic Virus and Tobacco Mosaic Virus Protein. Biochemistry 6, 3076-3081 (1967).

8. LAUFFER, M.A.: Tobacco Mosaic Virus and Its Protein. In: Subunits in Biological Systems (S.N.TIMASHEFF, G.E.FASMAN, eds.), part A, chap.4, pp.149-199. New York: Dekker 1971.

9. JAENICKE R., LAUFFER, M.A.: Determination of Hydration and Partial Specific Volume of Proteins with the Spring-Balance. Biochemistry 8, 3077-3082 (1969).

10. SHALABY, R.A.F., BANERJEE, K., LAUFFER, M.A.: Ion Binding by Tobacco Mosaic Virus and Its Protein. Biochemisty 7, 955-960 (1968).

11. BANERJEE, K., LAUFFER, M.A.: Role of Cation and Anion in the Polymerization and Depolymerization of Tobacco Mosaic Virus Protein. Biochemistry 10, 1100-1102 (1971).

12. FRANK, H.S., EVANS, M.W.: Free Volume and Entropy in Condensed Systems. III. Entropy in Binary Liquid Mixtures; Partial Molar Entropy in Dilute Solutions; Structure and Thermodynamics in Aqueous Electrolytes. J.Chem. Phys. 13, 507-532 (1945).

13. SMITH, C.E., LAUFFER, M.A.: Polymerization-Depolymerization of Tobacco Mosaic Virus Protein. VIII. Light Scattering Studies. Biochemistry 6, 2457-2465 (1967).

14. SHALABY, R.A.F., LAUFFER, M.A.: Polymerization-Depolymerization of Tobacco Mosaic Virus Protein. IX. Effect of Various Chemicals. Biochemistry 6, 2465-2473 (1967).

15. KHALIL, M.T.M., LAUFFER, M.A.: Polymerization-Depolymerization of Tobacco Mosaic Virus Protein. X. Effect of D_2O. Biochemistry 6, 2474-2480 (1967).

16. GORTNER, R.A.: The State of Water in Colloidal and Living Systems. Trans. Faraday Soc. 26, 678-686 (1930).

Chapter 3

1. STANLEY, W.M.: Isolation of a Crystalline Protein Possessing the Properties of Tobacco Mosaic Virus. Science 81, 644-645 (1935).

2. BAWDEN, F.C., PIRIE, N.W.: The Isolation and Some Properties of Liquid Crystalline Substances from Solanaceous Plants Infected with Three Strains of Tobacco Mosaic Virus. Proc.Roy.Soc.B. 123, 274-320 (1937).

3. STANLEY, W.M.: Virus Proteins - A New Group of Macro-Molecules. J.Phys. Chem. 42, 55-70 (1938).

4. LORING, H.S.: Properties of Hydrolytic Products of Nucleic Acid from Tobacco Mosaic Virus. J.Biol.Chem. 130, 251-258 (1939).

5. ERIKSSON-QUENSEL, I.B., SVEDBERG, T.: Sedimentation and Electrophoresis of the Tobacco Mosaic Virus Protein. J.Am.Chem.Soc. 58, 1863-1867 (1936).

6. WYCKOFF, R.W.F.: Molecular Sedimentation Constants of Tobacco Mosaic Virus Proteins Extracted from Plants at Intervals after Inoculation. J.Biol. Chem. 121, 219-224 (1937).

7. TAKASAHI, W.N., RAWLINS, T.E.: Methods for Determining Shape of Colloidal Particles: Application in Study of Tobacco Mosaic Virus. Proc.Soc.Exptl. Biol.Med. 30, 155 (1932).

8. LAUFFER, M.A., STANLEY, W.M.: Stream Double Refraction of Virus Proteins. J.Biol.Chem. 123, 507 (1938).

9. LAUFFER, M.A.: The Molecular Weight and Shape of Tobacco Mosaic Virus Protein. Science 87, 469-470 (1938).

10. WATSON, J.D.: The Structure of Tobacco Mosaic Virus. I. X-ray Evidence of a Helical Arrangement of Sub-units Around the Longitudinal Axis. Biochim. Biophys.Acta.13, 10-19 (1954)

11. FRANKLIN, R.E., HOLMES, K.C.: Tobacco Mosaic Virus: Application of the Method of Isomorphous Replacement to the Determination of the Helical Parameters and Radial Density Distribution. Acta.Cryst.11, 213-220 (1958).

12. SCHRAMM, G., SCHUMACHER, G., ZILLIG, W.: Über die Struktur des Tabakmosaikvirus. Z.Naturforsch.10b, 481-492 (1955).

13. HART, R.G.: Electron-microscopic Evidence for the Localization of Ribonucleic Acid in the Particles of Tobacco Mosaic Virus. Proc.Natl.Acad.Sci. U.S.41, 261-264 (1955).

14. MARKHAM, R., HITCHBORN, J.H., HILLS, G.J., FREY, S.: The Anatomy of the Tobacco Mosaic Virus. Virology 22, 342-359 (1964).

15. LAUFFER, M.A., STEVENS, C.L.: Structure of the Tobacco Mosaic Virus Particle; Polymerization of Tobacco Mosaic Virus Protein. Advan.Virus Res.13, 1-63 (1968).

16. SCHRAMM, G.: Über die Spaltung des Tabakmosaikvirus und die Wiedervereinigung der Spaltstücke zu höhermolekularen Proteinen, II Versuche zur Wiedervereinigung der Spaltstücke. Z.Naturforsch.2b, 249-257 (1947).

17. FRAENKEL-CONRAT, H.: Degradation of Tobacco Mosaic Virus with Acetic Acid. Virology 4, 1-4 (1957).

18. ANDERER, F.A., WITTMANN-LEIBOLD, B., WITTMANN, H.G.: Weitere Untersuchungen zur Aminosäuresequenz des Proteins in Tabakmosaikvirus. Z.Naturforsch. 20b, 1203-1213 (1965).

19. BANERJEE, K., LAUFFER, M.A.: Polymerization-Depolymerization of Tobacco Mosaic Virus Protein. VI. Osmotic Pressure Studies of Early Stages of Polymerization. Biochemistry 5, 1957-1963 (1966).

20. CASPAR, D.L.D.: Assembly and Stability of the Tobacco Mosaic Virus Particle. Advan.Protein Chem.18, 37-121 (1963).

21. ANSEVIN, A.T., LAUFFER, M.A.: Native Tobacco Mosaic Virus Protein of Molecular Weight 18,000. Nature 183, 1601-1602 (1959).

22. DURHAM, A.C.H., KLUG, A.: Polymerization of Tobacco Mosaic Virus and its Control. Nature New Biology 229, 42-46 (1971).

23. ANDERER, F.A., KRATKY, O., LO, R.: Untersuchung des gelösten Tabakmosaikvirus-Proteins mittels der Röntgen-Kleinwinkelstreuung unter Verwendung der Absolutintensität. Z.Naturforsch.10b, 906-916 (1964).

24. ANDERER, F.A.: Das Molekulargewicht der Peptideinheit im Protein des Tabakmosaikvirus. Z.Naturforsch.14b, 24-28 (1959).

25. WITTMANN, H.G.: Darstellung und Physiko-chemische Charakterisierung der Peptidketten der Tabakmosaikvirus. Experientia 15, 174-175 (1959).

26. FRIST, R.H., BENDET, I.J., SMITH, K.M., LAUFFER, M.A.: The Protein Subunit of Cucumber Virus 4; Degradation of Viruses by Succinylation. Virology 26, 558-566 (1965).

27. LAUFFER, M.A., SHALABY, R.A.F., KHALIL, M.T.M.: Polymerization-Depolymerization of Tobacco Mosaic Virus Protein: Refinement of a Model. Chimia (Aarau) 21, 460-462 (1967).

28. PAGLINI, S., LAUFFER, M.A.: Polymerization-Depolymerization of Tobacco Mosaic Virus Protein. XI. Osmotic Pressure Studies of Solutions in Water and in Heavy Water. Biochemistry 7, 1827-1834 (1968).

29. MARKHAM, R., FREY, S., HILLS, G.J.: Methods for the Enhancement of Image Detail and Accentuation of Structure in Electron Microscopy. Virology 20, 88-102 (1963).

30. FRANKLIN, R.E., COMMONER, B.: X-ray Diffraction by an Abnormal Protein (B-8) Associated with Tobacco Mosaic Virus. Nature 175, 1076-1077 (1955).

31. FINCH, J.T., LEBERMAN, R., YU-SHANG, C., KLUG, A.: Rotational Symmetry of the Two Turn Disk Aggregate of Tobacco Mosaic Virus Protein. Nature 212, 349-350 (1966).

32. FLORY, P.J.: Molecular Size Distribution in Linear Condensation Polymers. J.Am.Chem.Soc.58, 1877-1885 (1936).

33. LAUFFER, M.A.: Tobacco Mosaic Virus and Its Protein. In: Subunits in Biological Systems (S.N.TIMASHEFF, G.E.FASMAN, eds.), part A, chap.4, pp 149-199. New York: Dekker 1971

34. DURHAM, A.C.H., KLUG, A.: Structure and Roles of the Polymorphic Forms of Tobacco Mosaic Virus Protein. III. A Model for the Association of A-protein into Disks. J.Mol.Biol.67, 315-332 (1972).

35. STAUFFER, H., SRINIVASAN, S., LAUFFER, M.A.: Calorimetric Studies on Polymerization-Depolymerization of Tobacco Mosaic Virus Protein. Biochemistry 9, 193-200 (1970).

36. SCHEELE, R.B., LAUFFER, M.A.: Acid-base Titrations of Tobacco Mosaic Virus Protein. Biochemistry 6, 3076-3081 (1967).

37. VOGEL, D., JAENICKE, R.: Conformational Changes and Proton Uptake in the Reversible Aggregation of Tobacco Mosaic Virus Protein. Eur.J.Biochem.41, 607-615 (1974).

38. DURHAM, A.C.H., FINCH, J.T., KLUG, A.: Stages of Aggregation of Tobacco Mosaic Virus Protein. Nature New Biology 229, 130-134 (1968).

39. DURHAM, A.C.H.: Structures and Roles of the Polymorphic Forms of Tobacco Mosaic Virus Protein. I. Sedimentation Studies. J.Mol.Biol.67, 289-305 (1972).

40. DeROSIER, D.J., KLUG, A.: Reconstitution of Three Dimensional Structures from Electron Micrographs. Nature 217, 130-134 (1968).

41. FINCH, J.T., KLUG, A.: Three-Dimensional Reconstruction of the Stacked-Disk Aggregate of Tobacco Mosaic Virus Protein from Electronmicrographs. Phil.Trans.Roy.Soc.B261, 211-219 (1971).

42. LAUFFER, M.A.: Endothermic Polymerization in Biological and Biochemical Processes; Polymerization of Tobacco Mosaic Virus Protein. Chimia (Aarau) 20, 89-90 (1966).

Chapter 4

1. LAUFFER, M.A.: Polymerization-Depolymerization of Tobacco Mosaic Virus Protein. The Molecular Basis of Neoplasia (A Collection of Papers Presented at the Fifteenth Annual Symposium on Fundamental Cancer Research, 1961, at the University of Texas M.D. Anderson Hospital and Tumor Institute at Houston, Houston, Texas) Austin: University of Texas Press, p. 192, 1962.

2. SMITH, C.E., LAUFFER, M.A.: Polymerization-Depolymerization of Tobacco Mosaic Virus Protein. VIII. Light Scattering Studies. Biochemistry 6, 2457-2465 (1967).

3. BANERJEE, K., LAUFFER, M.A.: Polymerization-Depolymerization of Tobacco Mosaic Virus Protein. VI. Osmotic Pressure Studies of Early Stages of Polymerization. Biochemistry 5, 1957-1964 (1966).

4. LAUFFER, M.A.: Tobacco Mosaic Virus and Its Protein. In: Subunits in Biological Systems (S.N.TIMASHEFF, G.E.FASMAN, eds.), Part A, chap.4, pp.149-199. New York: Dekker 1971.

5. LAUFFER, M.A., ANSEVIN, A.T., CARTWRIGHT, T.E., BRINTON, C.C., Jr.: Polymerization-Depolymerization of Tobacco Mosaic Virus Protein. Nature 181, 1338-1339 (1958).

6. EISKAMP, J.G.: A Study of the Titrimetric, Sedimentation and Optical Behavior of the Structural Protein of Tobacco Mosaic Virus in Various States of Aggregation. Ph.D. Dissertation. Eugene/Oregon: University of Oregon 1969.

7. CASPAR, D.L.D.: Assembly and Stability of the Tobacco Mosaic Virus Particle. Advan.Protein Chem.18, 37-121 (1963).

8. PAGLINI, S., LAUFFER, M.A.: Polymerization-Depolymerization of Tobacco Mosaic Virus Protein. XI. Osmotic Pressure Studies of Solutions in Water and in Heavy Water. Biochemistry 7, 1827-1835 (1968).

9. DURHAM, A.C.H., FINCH, J.T., KLUG, A.: States of Aggregation of Tobacco Mosaic Virus Protein. Nature New Biology 229, 37-42 (1971).

10. DURHAM, A.C.H., FINCH, J.T.: Structures and Roles of the Polymorphic Forms of Tobacco Mosaic Virus Protein II. Electron Microscope Observations of the Larger Polymers. J.Mol.Biol.67, 307-314 (1972).

11. WESTOVER, C.J., Jr.: Sedimentation Equilibrium Studies on TMV Protein. Ph.D.Dissertation. Pittsburgh/Pa.: University of Pittsburgh 1971.

12. TANIGUCHI, M.: The Trimer-Disc Transformation of Tobacco Mosaic Virus Protein. Biochim.Biophys.Acta 181, 244-249 (1969).

13. SRINIVASAN, S., LAUFFER, M.A.: Polymerization Studies on Protein From Dahlemense Strain of TMV: Light Scattering and Related Studies. Arch. Biochem.Biophys.158, 53-66 (1973).

14. STEVENS, C.L.: The Condensation-Nucleation-Condensation Model for the Polymerization of Tobacco Mosaic Virus Protein. (manuscript in preparation, 1974).

15. ANSEVIN, A.T.F., LAUFFER, M.A.: Polymerization-Depolymerization of Tobacco Mosaic Virus Protein, I. Kinetics. Biophys.J.3, 239-251 (1963).

16. KHALIL, M.T.M., LAUFFER, M.A.: Polymerization-Depolymerization of Tobacco Mosaic Virus Protein. X. Effect of D_2O. Biochemistry 6, 2474-2480 (1967).

17. SHALABY, R.A.F., LAUFFER, M.A.: Polymerization-Depolymerization of Tobacco Mosaic Virus Protein. IX. Effect of Various Chemicals. Biochemistry 6, 2465-2473 (1967).

18. LAUFFER, M.A., STEVENS, C.L.: Structure of the Tobacco Mosaic Virus Particle; Polymerization of Tobacco Mosaic Virus Protein. Advan.Virus Res.13, 1-63 (1968).

19. ANSEVIN, A.T., STEVENS, C.L., LAUFFER, M.A.: Polymerization-Depolymerization of Tobacco Mosaic Virus Protein. III. Changes in Ionization and in Electrophoretic Mobility. Biochemistry 3, 1512-1518 (1964).

20. SCHEELE, R.B., LAUFFER, M.A.: Acid-Base Titrations of Tobacco Mosaic Virus and Tobacco Mosaic Virus Protein. Biochemistry 6, 3076-3081 (1967).

21. GORIN, M.H., in: Electrophoresis of Proteins and the Chemistry of Cell Surfaces (H.A.ABRAMSON, L.S.MOYER, M.H.GORIN) p.126. New York: RHEINHOLD 1942.

22. HILL, T.L.: Approximate Calculation of the Electrostatic Free Energy of Nucleic Acids and Other Cylindrical Macromolecules. Arch.Biochem.Biophys. 57, 229-239 (1955).

23. SRINIVASAN, S., LAUFFER, M.A.: Reconstitution of Tobacco Mosaic Virus: Calorimetric and Related Studies. Biochemistry 9, 2173-2180 (1970).

24. JAENICKE, R., LAUFFER, M.A.: Polymerization-Depolymerization of Tobacco Mosaic Virus Protein, XII. Further Studies on the Role of Water. Biochemistry 8, 3083-3092 (1969).

25. STAUFFER, H., SRINIVASAN, S., LAUFFER, M.A.: Calorimetric Studies on Polymerization-Depolymerization of Tobacco Mosaic Virus Protein. Biochemistry 9, 193-200 (1970).

Chapter 5

1. BAWDEN, F.C., PIRIE, N.W.: The Relationships Between Liquid Crystalline Preparations of Cucumber Viruses 3 and 4 and Strains of Tobacco Mosaic Virus. Brit.J.Exptl.Pathol.18, 275-291 (1937).

2. KNIGHT, C.A. and STANLEY, W.M.: Preparation and Properties of Cucumber Virus 4. J.Biol.Chem.141, 29-38 (1941).

3. KNIGHT, C.A.: Are Cucumber Viruses 3 and 4 Strains of Tobacco Mosaic Virus? A Review of the Problem. Virology 1, 216-267 (1955).

4. FRIST, R.H., BENDET, I.J., SMITH, K.M., LAUFFER, M.A.: The Protein Subunit of Cucumber Virus 4; Degradation of Viruses by Succinylation. Virology 26, 558-566 (1965).

5. KNIGHT, C.A.: The Nature of Some of the Chemical Differences Among Strains of Tobacco Mosaic Virus. J.Biol.Chem.171, 297-308 (1947).

6. TUNG, J.-S., KNIGHT, C.A.,: The Coat Protein Subunits of Cucumber Viruses 3 and 4 and a Comparison of Methods for Determining Their Molecular Weights," Virology 48, 574-581 (1972).

7. FRIST, R.H.: A Study of Cucumber Virus 4 Protein. Ph.D. Dissertation. Pittsburgh/Pa.: University of Pittsburgh 1965.

8. KRAMER, E., WITTMANN, H.G.: Elektrophoretische Untersuchungen der A-Proteine dreier Tabakmosaikvirus-Stämme. Z.Naturforsch.13b, 30-33 (1958).

9. MELCHERS, G.: Die biologische Untersuchung des Tomatenvirus Dahlem, 1940. Biol.Zbl.60, 527-537 (1940).

10. WITTMANN-LIEBOLD, B., WITTMANN, H.G.: Die primäre Proteinstruktur von Stämmen des Tabakmosaikvirus. Z.Vererbungslehre 94, 427-435 (1963).

11. SARKAR, S.: Evidence of Phenotipic Mixing Between Two Strains of Tobacco Mosaic Virus. Mol.Gen.Genetics 105, 87-90 (1969).

12. CASPAR, D.L.D., HOLMES, K.C.: Structure of Dahlemense Strain of Tobacco Mosaic Virus; A Periodically Deformed Helix. J.Mol.Biol.46, 99-133 (1969).

13. SRINIVASAN, S., LAUFFER, M.A.: Polymerization Studies on Protein from Dahlemense Strain of TMV: Light Scattering and Related Studies. Arch. Biochem.Biophys.158, 53-66 (1973).

14. OEHLEN, G.: A Comparison of the Number of Titratable Carboxyl Groups of TMV Wild Strains and Mutants in the Native and Denatured States. Mol.Gen. Genetics 99, 372-375 (1967).

15. SCHEELE, R.B., LAUFFER, M.A.: Acid-base Titrations of Tobacco Mosaic Virus and Tobacco Mosaic Virus Protein. Biochemistry 6, 3076-3081 (1967).

16. FRAENKEL-CONRAT, H., RAMACHANDRAN, K.L.: Structural Aspects of Tobacco Mosaic Virus. Advan.Protein Chem. 14, 175-229 (1959).

17. CASPAR, D.L.D.: Assembly and Stability of the Tobacco Mosaic Virus Particle. Advan.Protein Chem. 18, 37-121 (1963).

18. SRINIVASAN, S., LAUFFER, M.A.: Polymerization Studies on Protein from the PM2 Strain of TMV: Light Scattering and Related Studies. Arch.Biochem. Biophys. 158, 67-74 (1973).

19. SIEGEL, A., ZAITLIN, M., SEHGAL, O.P.: The Isolation of Defective Tobacco Mosaic Virus Strains. Proc.Natl.Acad.Sci.U.S. 48, 1845-1851 (1962).

20. ZAITLIN, M., McCAUGHLEHY, W.F.: Amino Acid Composition of a Nonfunctional Tobacco Mosaic Virus Protein. Virology 26, 500-503 (1965).

21. WITTMANN, H.G.: Die Proteinstruktur der Defektmutante PM2 des Tabakmosaikvirus. Z.Vererbungslehre 97, 297-304 (1965).

22. SEIGEL, A., HILLS, G.J., MARKHAM, R.: *In vitro* and *in vivo* Aggregation of the Defective PM2 Tobacco Mosaic Virus Protein. J.Mol.Biol. 19, 140-144 (1966).

23. SMITH, C.E., LAUFFER, M.A.: Polymerization-Depolymerization of Tobacco Mosaic Virus Protein. VIII. Light Scattering Studies. Biochemistry 6, 2457-2465 (1967).

24. OKADA, Y., OHASHI, Y., OHNO, T.E., NOZU, Y.: Sequential Reconstitution of Tobacco Mosaic Virus. Virology 42, 243-245 (1970).

25. OHNO, T., INOUE, H., OKADA, Y.: Assembly of Rod-Shaped Virus *in vitro*: Reconstitution with Cucumber Green Mottle Mosaic Virus Protein and Tobaccc Mosaic Virus RNA. Proc.Nat.Acad.Sci.U.S. 69, 3680-3683 (1972).

26. LISTER, R.M.: Possible Relationship of Virus-Specific Products of Tobacco Rattle Virus Infections. Virology 28, 350-353 (1966).

27. MORRIS, J.T., SEMANCIK, J.S.: *In vitro* Protein Polymerization and Nucleoprotein Reconstitution of Tobacco Rattle Virus. Virology 53, 215-224 (1973).

28. FRITSCH, C., WITZ, J., ABOU HAIDAR, M., HIRTH, L.: Polymerization of Tobacco Rattle Virus Protein. FEBS Letters 29, 211-214 (1973).

29. SEMANCIK, J.S.: Identity of Structural Proteins from Two Isolates of TRV with Different Lengths of Associated Short Particles. Virology 40, 618-623 (1970).

30. LAUFFER, M.A.: Polymerization-Depolymerization of Tobacco Mosaic Virus Protein," Proceedings of the 15th Annual Symposium on Fundamental Cancer Research at the University of Texas M.D. Anderson Hospital and Tumor Institute, University Texas Press, Austin, Texas, 1962, pp. 180-206.

31. BANERJEE, K., LAUFFER, M.A.: Polymerization-Depolymerization of Tobacco Mosaic Virus Protein, VI. Osmotic Pressure Studies of Early Stages of Polymerization. Biochemistry 5, 1957-1964 (1966).

32. SCHRAMM, G., ZILLIG, W.: Über die Struktur des Tabakmosaikvirus, IV. Mitt: Die Reaggregation des nucleinsäurefreien Proteins. Z.Naturforsch. 106, 493-499 (1955).

33. STEVENS, C.L., LAUFFER, M.A.: Polymerization-Depolymerization of Tobacco Mosaic Virus Protein. IV. The Role of Water. Biochemistry 4, 31-37 (1965).

34. JAENICKE, R., LAUFFER, M.A.: Polymerization-Depolymerization of Tobacco Mosaic Virus Protein. XII. Further Studies on the Role of Water. Biochemistry 8, 3083-3092 (1969).

35. ANSEVIN, A.T., STEVENS, C.L., LAUFFER, M.A.: Polymerization-Depolymerization of Tobacco Mosaic Virus Protein. III. Changes in Ionization and in Electrophoretic Mobility. Biochemistry 3, 1512-1518 (1964).

36. LAUFFER, M.A.: Polymerization-Depolymerization of Tobacco Mosaic Virus Protein. VII. A model. Biochemistry 5, 2440-2446 (1966).

37. LAUFFER, M.A.: Tobacco Mosaic Virus and its Protein. In: Biological Systems (S.N.TIMASHEFF, F.E.FASMAN, eds.), part A, chap.4, pp.149-199. New York: Dekker 1971.

38. BULL, H.B.: Adsorption of Water Vapor by Protein. J.Am.Chem.Soc.66, 1499-1507 (1944).

39. McLAREN, A.D., ROWEN, J.W.: Sorption of Water Vapor by Proteins and Polymers: A review. J.Polymer Sci.7, 289-324 (1951).

40. DURHAM, A.C.H., FINCH, J.T., KLUG, A.: States of Aggregation of Tobacco Mosaic Virus Protein. Nature New Biology 229, 37-42 (1971).

41. LAUFFER, M.A.: Endothermic Polymerization in Biological and Biochemical Processes: Polymerization of Tobacco Mosaic Virus Protein. Chimia (Aarau) 20, 89-90 (1966).

42. DURHAM, A.C.H., KLUG, A.: Polymerization of Tobacco Mosaic Virus and its Control. Nature New Biology 229, 42-46 (1971).

43. ANSEVIN, A.T., LAUFFER, M.A.: Polymerization-Depolymerization of Tobacco Mosaic Virus Protein I. Kinetics. Biophys.J.3, 239-251 (1963).

44. LAUFFER, M.A., SHALABY, R.A.F., KHALIL, M.T.M.: Polymerization-Depolymerization of Tobacco Mosaic Virus Protein: Refinement of a Model. Chimia (Aarau) 21, 460-462 (1967).

45. SCHUSTER, T.M., SCHEELE, R.B.: Mechanism of Self Assembly of Tobacco Mosaic Virus Protein. III. Temperature - Jump Relaxation Kinetics. (manuscript in preparation). (1974).

46. SRINIVASAN, S., LAUFFER, M.A.: Reconstitution of Tobacco Mosaic Virus Protein: Calorimetric and Related Studies. Biochemistry 9, 2173-2180 (1970).

47. JANEICKE, R., LAUFFER, M.A.: Polymerization-Depolymerization of Tobacco Mosaic Virus Protein. XII. Further Studies on the Role of Water. Biochemistry 8, 3083-3092 (1969).

48. WESTOVER, C.J., Jr.: Sedimentation Equilibrium Studies of TMV Protein. Ph.D. Dissertation. Pittsburgh/Pa.: University of Pittsburgh 1971.

49. von SENGBUSCH, P.: Influence of Protein Structure on Selection of Nitrous Acid Induced Mutants of TMV. Mol.Gen.Genetics 99, 171-180 (1967).

Chapter 6

1. SCHRAMM, G.: Über die Spaltung des Tabakmosaikvirus in niedermolekulare Proteine und die Rückbildung hochmolekularen Proteins aus den Spaltstükken. Naturwissenschaften 31, 94-96 (1943).
2. SCHRAMM, G.: Über die Spaltung des Tabakmosaikvirus und die Wiedervereinigung der Spaltstücke zu höhermolekularen Proteinen. II. Versuche zur Wiedervereinigung der Spaltstücke. Z.Naturforsch.2b, 249-257 (1943).
3. FRAENKEL-CONRAT, H., WILLIAMS, R.C.: Reconstitution of Active Tobacco Mosaic Virus from its Inactive Protein and Nucleic Acid Components. Proc. Natl.Acad.Sci.U.S.41, 690-698 (1955).
4. LIPPINCOTT, J.A., COMMONER, B.: Reactivation of Tobacco Mosaic Virus Infectivity in Mixtures of Virus Protein and Nucleic Acid. Biochim.Biophys. Acta 19, 198-199 (1956).
5. FRAENKEL-CONRAT, H.: The Role of the Nucleic Acid in the Reconstitution of Active Tobacco Mosaic Virus. J.Am.Chem.Soc.78, 882-883 (1956).
6. GIERER, A., SCHRAMM, G.: Infectivity of Ribonucleic Acid from Tobacco Mosaic Virus. Nature 117, 702-703 (1956).
7. LAUFFER, M.A., TRKULA, D., BUZZELL, A.: Mechanism of Inactivation of Tobacco Mosaic Virus by X-rays. Nature 117, 890 (1956).
8. FRAENKEL-CONRAT, H., SINGER, B.: Reconstitution of Tobacco Mosaic Virus. IV. Inhibition by Enzymes and Other Proteins, and use of Polynucleotides. Virology 23, 354-362 (1964).
9. FRAENKEL-CONRAT, H., SINGER, B.: Virus Reconstitution. II. Combination of Protein and Nucleic Acid from Different Strains. Biochem.Biophys.Acta 24, 540-548 (1957).
10. HART, R.G., SMITH, J.D.: Interactions of Ribonucleotide Polymers with Tobacco Mosaic Virus Protein to form Virus-like Particles. Nature 178, 739-740 (1958).
11. HOLOUBEK, V.: Mixed Reconstitution Between Protein from Common Tobacco Mosaic Virus and Ribonucleic Acid from other Strains. Virology 18, 401-404 (1962).
12. TSUTOMU SUGIYAMA: Tobacco Mosaic Virus-like Rods Formed by Mixed Reconstitution between MS2 Ribonucleic Acid and Tobacco Mosaic Virus Protein. Virology 28, 488-492 (1966).
13. MATTHEWS, R.E.F., HARDIE, J.D.: Reconstitution of RNA from Spherical Viruses with Tobacco Mosaic Virus Protein. Virology 28, 165-168 (1966).
14. HIEBERT, E., BANCROFT, J.B., BRACKER, C.E.: The Assembly *in Vitro* of some Small Spherical Virus, Hybrid Viruses, and other Nucleoprotein. Virology 34, 492-508 (1968).
15. VERDUIN, B.J.M., BANCROFT, J.B.: The Infectivity of Tobacco Mosaic Virus RNA in Coat Proteins from Spherical Viruses. Virology 37, 501-506 (1969).
16. REDDI, K.K.: Studies on the Formation of Tobacco Mosaic Virus with Emphasis on the Events Within the Host Cell Following Infection. Advan. Virus Res.17, 51-94 (1972).
17. HIRAI, T., WILDMAN, S.G.: Cytological and Cytochemical Observations on the Early Stage of Infection of Tomato Hair Cells by Tobacco Mosaic Virus. Plant Cell Physiol.(Tokyo) 4, 265-275 (1963).

18. BALD, J.G.: Cytological Evidence for the Production of Plant Virus Ribonucleic Acid in the Nucleus. Virology 22, 377-387 (1964).

19. YASUDA, Y., HIRAI, T.: Incorporation of ^{3}H-uracil into Tobacco Leaf Epidermis Infected with Tobacco Mosaic Virus. Exptl.Cell Res.34, 210-212 (1964).

20. SMITH, S.H., SCHLEGEL, D.E.: The Incorporation of Ribonucleic Acid Precursors in Healthy and Virus-Infected Plant Cells. Virology 26, 180-189 (1956).

21. NILSSON-TILLGREN, T.: Studies on the Biosynthesis of TMV III. Isolation and Characterization of the Replicative form and the Replicative Intermediate RNA. Mol.Gen.Genetics 109, 246-256 (1970).

22. SHALLA, E.A., AMICI, A.: The Distrubution of Viral Antigen in Cells Infected with Tobacco Mosaic Virus as Revealed by Electron Microscopy. Virology 31, 78-91 (1964).

23. HIRAI, A., WILDMAN, S.G.: Intracellular Site of Assembly of TMV-RNA and Protein. Virology 33, 467-473 (1967).
Proc.Natl.Acad.Sci.U.S.52, 398-401 (1964).

24. SINGER, B.: Protein Synthesis in Virus-Infected Plants II. The Synthesis and Cumulation of TMV and TMV Coat Protein in Subcellular Fractions of TMV-Infected Tobacco. Virology 47, 397-404 (1972).

25. ESAU, K., CRONSHAW, G.: Relation of Tobacco Mosaic Virus to the Host Cells. G.Cell Biol.33, 665-678 (1967).

26. SRINIVASAN, S., LAUFFER, M.A.: Reconstitution of Tobacco Mosaic Virus: Calorimetric and Related Studies. Biochemistry 9, 2173-2180 (1970).

27. STAUFFER, H., SRINIVASAN, S., LAUFFER, M.A.: Calorimetric Studies on Polymerization-Depolymerization of Tobacco Mosaic Virus Protein. Biochemistry 9, 193-200 (1970).

28. LAUFFER, M.A.: The Influence of Concentration Upon the Sedimentation Rate of Tobacco Mosaic Virus. J.Am.Chem.Soc.66, 1195-1201 (1944).

29. SRINIVASAN, S., LAUFFER, M.A.: Conformational Studies of TMV-RNA: Calorimetric and Related Investigations. Arch.Biochem.Biophys.148, 568-575 (1971).

30. CANTOR, C.R., JASKUNAS, S.R., TINOCO, I.: Optical Properties of Ribonucleic Acids Predicted from Oligomers. J.Mol.Biol.20, 39-62 (1966).

31. KAUZMANN, W.: Some Factors in the Interpretation of Protein Denaturation. Advan.Protein Chem.14, 1-63 (1959).

32. SCHEELE, R.B., LAUFFER, M.A.: Acid-Base Titrations of Tobacco Mosaic Virus and Tobacco Mosaic Virus Protein. Biochemistry 6, 3076-3081 (1967).

33. COX, R.A., JONES, A.S., MARSH, C.E., PEACOCKE, A.R.: On Hydrogen Bonding and Branching in a Bacterial Ribonucleic Acid. Biochem.Biophys.Acta 21, 576-577 (1956).

34. KRAMER, E., WITTMANN, H.G.: Elektrophoretische Untersuchungen der A-Proteine dreier Tabakmosaikvirus-Stämme. Z.Naturforsch.13b, 30-33 (1958).

35. STEVENSON, D.A., LAUFFER, M.A.: unpublished results (1974).

36. ADIARTE, A.L., LAUFFER, M.A.: The Donnan Effect in Tobacco Mosaic Virus and its Components. Arch.Biochem.Biophys.158, 75-83 (1973).

37. BUTLER, P.J.G., KLUG, A.: Assembly of the Particle of Tobacco Mosaic Virus from RNA and Discs of Protein. Nature New Biology 229, 47-50 (1971).

38. BUTLER, P.J.G.: Structures and Roles of the Polymorphic Forms of Tobacco Mosaic Virus Protein VI. Assembly of the Nucleoprotein Rods of Tobacco Mosaic Virus from the Protein Disks and RNA. J.Mol.Biol.72, 25-35 (1972).

39. DURHAM, A.C.H., FINCH, J.T., KLUG, A.: Stages of Aggregation of Tobacco Mosaic Virus Protein. Nature New Biology 229, 37-42 (1971).

40. DURHAM, A.C.H., KLUG, A.: Polymerization of Tobacco Mosaic Virus Protein and its Control. Nature New Biology 229, 42-46 (1971).

41. STUSSI, C., LEBEURIER, G., HIRTH, L.: Partial Reconstitution of Tobacco Mosaic Virus. Virology 38, 16-25 (1969).

42. RODIONOVA, N.P., VESENINA, N.E., KICHATOVA, O.V., ATABEKOV, J.G.: An Intermediate Nucleoprotein Complex Formed on Tobacco Mosaic Virus Reconstitution. Virology 46, 183-191 (1971).

43. RODIONOVA, N.P., VESENINA, N.E., ATABEKOVA, T.I., DZHAVAKHIA, V.G., ATABEKOV, J.G.: Further Studies on the Reconstitution of TMV and an Incomplete Nucleoprotein Complex. Virology 51, 24-33 (1973).

44. GUILLEY, H., STUSSI, C., HIRTH, L.: Influence de la phosphorodiesterase de rate de Porc sur la reconstitution *in vitro* du Virus de la Mosaique du Tabac. C.R.Acad.Sc.Paris 272, 1181-1184 (1971).

45. THOUVENEL, J.C., GUILLEY, H., STUSSI, C., HIRTH, L.: Evidence for Polar Reconstitution of TMV. FEBS Letters 16, 204-206 (1971).

46. STUSSI, C., GUILLEY, H., LEBEURIER, G., HIRTH, L.: Some Recent Advances in the Comprehension of *in vitro* Morphogenesis of Tobacco Mosaic Virus. Biochimie 54, 287-296 (1972).

47. OHNO T., NOZU, Y., OKADA, Y.: Polar Reconstitution of Tobacco Mosaic Virus (TMV). Virology 44, 510-516 (1971).

48. MAY, D.S., KNIGHT, C.A.: Polar Stripping of Protein Subunits From Tobacco Mosaic Virus. Virology 25, 502-507 (1965).

49. RICHARDS, K.E., WILLIAMS, R.C.: Assembly of Tobacco Mosaic Virus *in vitro*: Effect of State of Polymerization of the Protein Component: Proc.Natl. Acad.Sci.U.S.69, 1121-1124 (1972).

50. OKADA, Y., OHNO, T.: Assembly Mechanism of Tobacco Mosaic Virus Particles from its Ribonucleic Acid and Protein. Mol.Gen.Genet.114, 205-213 (1972).

51. BUTLER, P.J.G., KLUG, A.: Assembly of Tobacco Mosaic Virus *in vitro*: Effect of State of Polymerization of the Protein Component. Proc.Natl. Acad.Sci.U.S.69, 2950-2953 (1972).

52. RICHARDS, K.E., WILLIAMS, R.C.: Assembly of Tobacco Mosaic Virus Rods *in vitro*: Elongation of Partially Assembled Rods. Biochemistry 12, 4574-4581 (1973).

53. OKADA, Y., OHNO, T., KURIYAMA, K., INOUE, H.: Participation of the Tyrosine Residue at Position 139 of Tobacco Mosaic Protein in the Reconstitution Reaction. Virology 47, 338-341 (1972).

54. OHNO, T., YAMAURA, R., KURIYAMA, K., INOUE, H., OKADA, Y.: Structure of N-Bromosuccinimide-Modified Tobacco Mosaic Virus Protein and its Function in the Reconstitution Process. Virology 50, 76-83 (1972).

55. OHNO, T., INOUE, H., OKADA, Y.: Assembly of Rod-Shaped Virus *in vitro*: Reconstitution with Cucumber Green Mottle Mosaic Virus Protein and Tobacco Mosaic Virus RNA. Proc.Natl.Acad.Sci.U.S.69, 3680-3683 (1972).

56. GUILLEY, H., JONARD, G., HIRTH, L.: A TMV-RNA Nucleotide Sequence Specifically Recognized by TMV Protein. Biochimie 56, 181-185 (1974).

57. MORRIS, J.T., SEMANCEK, J.S.: *in vitro* Protein Polymerization and Nucleoprotein Reconstitution of Tobacco Rattle Virus. Virology 53, 215-224 (1973).

58. HAIDAR, M.A., PFEIFFER, P., FRITSCH, C., HIRTH, L.: Sequential Reconstitution of Tobacco Rattle Virus. J.Gen.Virol.21, 83-91 (1973).

59. ATABEKOV, J.G., NOVIKOV, V.K., VISHNICHENKO, V.K., KAFTANOVA, A.S.: Some Properties of Hybrid Viruses Reassembled *in vitro*. Virology 41, 519-532 (1970).

Chapter 7

1. PAULING, L.: Abnormality of Hemoglobin Molecules. In: Hereditary Hemolytic Anemias. The Harvey Lectures, 1953-1954, series XLIX. pp.216-242. New York: Academic Press 1955.

2. BRAUNITZER, G., GEHRING-MÜLLER, R., HILSCHMANN, N., HILSE, K., HOBOM, G., RUDLOFF, V., WITTMANN-LIEBOLD, B.: Die Konstitution des normalen adulten Humanhämoglobins. Z.Physiol.Chem.325, 283-286 (1961).

3. KONIGSBERG, W., GUIDOTTI, G., HILL, R.J.: The Amino Acid Sequence of the Chain of Human Hemoglobins. J.Biol.Chem.236, 55-56 (1961).

4. KONIGSBERG, W., GOLDSTEIN, J., HILL, R.J.: The Structure of Human Hemoglobin. VII. The Digestion of the β Chain of Human Hemoglobin with Pepsin. J.Biol.Chem.238, 2016-2033 (1963).

5. INGRAM, V.M.: Gene Mutations in Human Haemoglobin: The Chemical Difference Between Normal and Sickle Cell Haemoglobin. Nature 180, 326-328 (1957).

6. HUNT, J.A., INGRAM, V.M.: A Terminal Peptide Sequence of Human Haemoglobin. Nature 184, 640-699 (1959).

7. HARRIS, J.W.: Studies on the Destruction of Red Blood Cells. VIII. Molecular Orientation in Physical Cell Hemoglobin Solution. Proc.Soc.Exptl. Biol.Med.75, 197-201 (1950).

8. MURAYAMA, M.: Molecular Mechanism of Red Cell Sickling. Science 153, 145-149 (1966).

9. MAGDOFF-FAIRCHILD, B., SWERDLOW, P.H., BERTLES, J.F.: Intermolecular Organization of Deoxygenated Sickle Haemoglobin Determined by X-ray Diffraction. Nature 239, 217-219 (1972).

10. FINCH, J.T., PERUTZ, M.F., BERTLES, J.F., DÖBLER, J.: Structure of Sickle Erythrocytes and of Sickle-Cell Hemoglobin Fibers. Proc.Natl.Acad.Sci. U.S.70, 718-722 (1973).

11. STETSON, C.A., Jr.: The State of Hemoglobin in Sickled Erythrocytes. J.Exptl.Med.123, 341-346 (1966).

12. DÖBLER, J., BERTLES, J.F.: The Physical State of Hemoglobin in Sickle-Cell Anemia Erythrocytes *in vivo*. J.Exptl.Med.127, 711-716 (1968).

13. MURAYAMA, M.: Titratable Sulfhydryl Groups of Normal and Sickle Cell Hemoglobins at 0° and 38°. J.Biol.Chem.228, 231-240 (1957).

14. MURAYAMA, M.: Thermal (or Endothermic) Aggregation of Sickle Cell Hemoglobin (Hb S) during Sickling. In: Hemoglobin and Red Cell Structure and Function (G.J.BREWER, ed.), pp.243-251. New York: Plenum Publishing Corp. 1972.

15. MURAYAMA, M.: Sickle Cell Hemoglobin: Molecular Basis of Sickling Phenomenon: Theory and Therapy. In: Critical Review of Biochemistry. Chemical Rubber, Co. 1973.

16. MURAYAMA, M.: A Molecular Mechanism of Sickled Erythrocyte Formation. Nature 202, 258-260 (1964).

17. PAULING, L., ITANO, H.A., SINGER, S.J., WELLS, I.C.: Sickle Cell Anemia, A Molecular Disease. Science 110, 543-548 (1949).

18. MURAYAMA, M.: A Sub-Molecular Mechanism of Gel Formation in Sickle Cell Haemolysate. Nature 194, 933-934 (1962).

19. THOMAS, L., STETSON, C.A., Jr.: Sulfhydryl Compounds and the Sickling Phenomenon. Bull.Johns Hopkins Hos. 83, 176-180 (1948).

20. MURAYAMA, M.: On the Nature of the Interaction Between Binding Sites for Heavy Metals (mercapto-mercapto interaction) in Normal Hemoglobin. J.Biol. Chem.234, 3158-3162 (1959).

21. RIGGS, A.F.: Sulfhydryl Groups and the Interaction Between the Hemes in Hemoglobin. J.Gen.Physiol.36, 1-16 (1952).

22. MURAYAMA, M., OLSON, R.A., JENNINGS, W.H.: Molecular Orientation in Horse Hemoglobin Crystals and Sickled Erythrocytes. Biochim.Biophys.Acta 94, 194-199 (1965).

23. MURAYAMA, M.: Orientation of Sickled Erythrocytes in a Magnetic Field. Nature 206, 420-422 (1965).

24. SIMMONS, N.S., BLOUT, E.R.: The Structure of Tobacco Mosaic Virus and Its Components; Ultraviolet Optical Rotatory Dispersion. Biophys.J.1, 55-62 (1960).

25. LAUFFER, M.A., SHALABY, R.A.F., KHALIL, M.T.M.: Polymerzation-Depolymerization of Tobacco Mosaic Virus Protein; Refinement of a Model. Chimia (Aarau) 21, 460-462 (1967).

26. JAENICKE, R., LAUFFER, M.A.: Polymerization-Depolymerization of Tobacco Mosaic Virus Protein. XII. Further Studies on the Role of Water. Biochemistry 8, 3083-3092 (1969).

27. MUIRHEAD, H., PERUTZ, M.A.: Structure of Haemoglobin: a Three-Dimensional Fourier Synthesis of Reduced Human Haemoglobin at 5.5 A° Resolution. Nature 199, 633-638 (1963).

28. WHITE, J.F., HEAGAN, B.: Gels of Normal and Sickled Hemoglobin. J.Exptl. Med.131, 1079-1092 (1970).

Chapter 8

1. GROSS, J., SCHMITT, F.O.: The Structure of Human Skin Collagen as Studied with the Electron Microscope. J.Exptl.Med.88, 555-568 (1948).

2. HARRINGTON, W.S., von HIPPEL, P.H.: The Structure of Collagen and Gelatin. Advan.Protein Chem.16, 1-138 (1961).

3. FESSLER, J.H.: Properties of Neutral-Salt-Soluable Collagen. I.Biochem. J.76, 463-471 (1960).

4. GROSS, J., KIRK, D.: The Heat Precipitation of Collagen from Neutral Salt Solution: Some Rate Regulating Factors. J.Biol.Chem.233, 355-360 (1958).

5. CASSEL, J.M., MANDELKERN, L., ROBERTS, D.E.: The Kinetics of the Heat Precipitation of Collagen. J.Am.Leather Chemists' Assoc.57, 556-575 (1962).

6. CASSEL, J.M., CHRISTENSEN, R.G.: Volume Change on Formation of Native Collagen Aggregate. Biopolymers 5, 431-437 (1967).

7. FESSLER, J.H., HODGE, A.J.: Ultracentrifical Observation of Phase Transitions in Density Gradients I. The Collagen System. J.Mol.Biol.5, 446-449 (1962).

8. FESSLER, J.H.: Collagen Fiber Formation Observed in Buoyant Density Gradients. In: Structure and Function of Connective and Skeletal Tissues (S.F.JACKSON, S.M.PARTRIDGE, R.D.HARKNESS, F.R.TRISTRAM, eds.), pp.80-86 London: Buttersworth 1965.

9. von MURALT, A., EDSALL, J.T.: Studies in the Physical Chemistry of Muscle Globulin II,III and IV. J.Biol.Chem.89, 289-386 (1930).

10. WEBER, H.H., STÖVER, R.: Das kolloidale Verhalten der Muskeleiweisskörper und das van der Wallssche Wirkungsvolumen der Myogenteilchen. Biochem. Z.259, 269-284 (1933).

11. LAUFFER, M.A., SZENT-GYÖRGYI, A.G.: Comments on the Structure of Myosin. Arch.Biochem.Biophys.56, 542-548 (1955).

12. LOWEY, S.: Myosin; Molecule and Filament. In: Subunits in Biological Systems (S.N.TIMASHEFF, G.E.FASMAN, eds.) part A, pp.201-259. New York: Dekker 1971.

13. LOWEY, S., SLAYTER, H.S., WEEDS, A.G., BAKER, H.: Substructure of the Myosin Molecule. I. Subfragments of Myosin by Enzymatic Degradation. J.Mol.Biol.42, 1-29 (1969).

14. HUXLEY, H.E.: The Mechanism of Muscular Contraction. Sci.Am.213, 18-27 (December, 1965).

15. GODFREY, J.E., HARRINGTON, W.F.: Self Association in the Myosin System at High Ionic Strength. II. Evidence for the Presence of a Monomer Dimer Equilibrium. Biochemistry 9, 894-908 (1970).

16. LAKI, K., CARROLL, W.R.: Size of the Myosin Molecule. Nature 175, 389-390 (1955).

17. KAMINER, B., BELL, A.L.: Myosin Filamentogenesis; Effects of *pH* and Ionic Concentration. J.Mol.Biol.20, 391-400 (1966).

18. JOSEPHS, R., HARRINGTON, W.F.: Studies on the Formation and Physical Chemical Properties of Synthetic Myosin Filaments. Biochemistry 5, 3474-3487 (1966).

19. JOSEPHS, R., HARRINGTON, W.F.: On the Stability of Myosin Filaments. Biochemistry 7, 2834-2847 (1968).

20. STRAUB, F.B.: Actin. Studies Inst.Med.Chem.Univ.Szeged.2, 3-15 (1943).

21. OOSAWA, F., KASAI, M.: Actin. In: Subunits in Biological Systems (S.N. TIMASHEFF, G.E.FASMAN, eds.), part A, pp.261-322. New York: Dekker 1971.

22. STRAUB, F.B., FEUER, G.: Adenosinetriphosphate, The Functional Group of Actin. Biochim.Biophys.Acta 4, 455-470 (1950).

23. HAYASHI, T., ROSENBLUTH, R.: Actin Polymerization by Direct Transphosphorylation. Biochem.Biophys.Res.Comm.8, 20-23 (1962).

24. GRANT, R.J.: Effect of Temperature on Polymerization of G-ADP-Actin. Federation Proc.22, 403 (1963).

25. IKKAI, T., OOI, T., NOGUCHI, H.: Actin: Volume Change on Transformation of G-form to F-form. Science 152, 1756-1757 (1966).

26. IKKAI, T., OOI, T.: The Effects of Pressure on F-G Transformation of Actin. Biochemistry 5, 1551-1560 (1966).

27. ASAKURA, S., OOSAWA, F.: Dephosphorylation of Adenosine Triphosphate in Actin Solutions at Low Concentrations of Magnesium. Arch.Biochem.Biophys. 87, 273-280 (1960).

28. SELBY, C., BEAR, R.: The Structure of Actin-Rich Filaments of Muscles According to X-Ray Diffraction. J.Biophys.Biochem.Cytol.2, 71-85 (1956).

29. HANSON, J., LOWY, J.: The Structure of F-Actin and of Actin Filaments Isolated From Muscles. J.Mol.Biol.6, 46-60 (1963).

30. KASAI, M., NAKANO, E., OOSAWA, F.: Polymerization of Actin Free From Nucleotides and Divalent Cations. Biochem.Biophys.Acta.94, 494-503 (1965).

31. OOSAWA, F., ASAKURA, S., HOTTA, K., IMAI, N., OOI, T.: G-F Transformation of Actin as a Fibrous Condensation. J.Polymer Sci.37, 323-336 (1959).

32. ASAKURA, S., KASAI, M., OOSAWA, F.: The Effect of Temperature on the Equilibrium State of Actin Solutions. J.Polymer Sci.44, 35-49 (1960).

33. KASAI, M.: Thermodynamical Aspects of G-F Transformations of Actin. Biochim.Biophys.Acta 180, 399-409 (1969).

34. SZENT-GYÖRGYI, A.: Thermodynamic Theory of the Contraction of Actomyosin. Biol.Bull.95, 285-286 (1948).

35. POLISSAR, M.J.: Physical Chemistry of Contractile Processes in Muscle. I. A Physiochemical Model of Contractile Mechanism. Am.J.Physiol.168, 766-781 (1952).

36. BROWN, D.E.S.: The Pressure-Tension-Temperature Relation in Cardiac Muscle. Am.J.Physiol.109, 16 (1934).

37. ASAKURA, S., EGUCHI, G., IINO, T.: Reconstitution of Bacterial Flagella *in vitro*. J.Mol.Biol.10, 42-56 (1964).

38. GERBER, B.R., ASAKURA, S., OOSAWA, F.: Effect of Temperature on the *in vitro* Assembly of Bacterial Flagella. J.Mol.Biol.74, 467-484 (1973).

39. GERBER, B.R., NOGUCHI, H.: Volume Change Associated with the G-F Transformation of Flagellin. J.Mol.Biol.26, 197-210 (1967).

40. GERBER, B.R., ROUTLEDGE, L.M., TAKASHIMA, S.: Self Assembly of Bacterial Flagellar Protein: Dielectric Behavior of Monomers and Polymers. J.Mol. Biol.71, 317-337 (1972).

Chapter 9

1. SZEGVARI, A.: Über Gelatinierung von Nitrozellulose Lösungen. Kolloid Z. 34, 34-37 (1924).

2. McBAIN, J.W., HARVEY, C.E., SMITH, L.E.: The Apparent Viscosity of Solutions of Nitro-cotton in Various Solvents. J.Phys.Chem.30, 312-352 (1926).

3. HEYMAN, E.: Studies on Sol-gel Transformations. I. The Inverse Sol-gel Transformation of Methylcellulose in Water. Trans.Faraday Soc.31, 846-864 (1935).

4. FREUNDLICH, H.: Some Recent Work on Gels. J.Phys.Chem.41, 901-910 (1937).

5. MARSLAND, D., BROWN, D.E.S.: The Effects of Pressure on Sol-gel Equilibria, with Special Reference to Myosin and other Protoplasmic Gels. J.Cellular Comp.Physiol.20, 295-305 (1942).

6. JOHNSON, F.H., EYRING, H., POLISSAR, M.J.: The Kinetic Basis of Molecular Biology. New York: Wiley 1954.

7. LAING, M.E.: Unstable States of Solutions of Sodium Behenate. J.Chem.Soc. (London) 127, 2751-2755 (1925).

8. DODDARD, E.D., HOEVE, C.A.J., BENSON, G.C.: Heats of Micelle Formation of Parafin Chain Salts in Water. J.Phys.Chem.61, 593-598 (1957).

9. DOTY, P., EPSTEIN, S.I.: Direct Observation of Hapten-Antibody Equilibria. Nature 174, 89-90 (1954).

10. HAUROWITZ, F., SOWINSKI, R., CHENG, H.F.: The Dissociation of Antigen-Antibody Precipitates. J.Am.Chem.Soc.79, 1882-1888 (1957).

11. SCOTT, E., BERNS, D.S.: Protein-Protein Interaction. The Phycocyanin System. Biochemistry 4, 2597-2606 (1965).

12. NOGUCHI, J., HAYAKAWA, T., EBATA, M.: Reversible Heat Coagulation of Some Water-Soluble Amino-Acid Copolymers. J.Polymer Sci.23, 843-849 (1956).

13. CIFERRI, A., ORIFINO, D.A.: Phase Separation of Poly-L-Proline in Salt Solutions. J.Phys.Chem.70, 3277-3285 (1966).

14. BROWN, B.L., JENNINGS, B.R.: Aggregation of Poly-L-Proline. II. Biopolymers 9, 1119-1124 (1970).

15. DOTY, P., YANG, J.T.: Polypeptides. VII. Poly-γ-benzyl-L-glutamate; the Helix-coil Transition in Solution. J.Am.Chem.Soc.78, 498-500 (1956).

16. SCHADE, A.L., REINHART, R.W.: Galactothermin, a Reversible Heat-Precipitable Protein of Human Milk at Neutral *pH*. Biochem.J.118, 181-186 (1970).

17. MURAYAMA, M.: Sickle Cell Hemoglobin: Molecular Basis of Sickling Phenomenon: Theory and Therapy. In: Critical Review of Biochemistry. Chemical Rubber Company, 1973.

18. KETTMAN, M.S., NISHIKAWA, A.H., MORITA, R.Y., BACKER, R.R.: Effect of Hydrostatic Pressure on the Aggregation Reaction of Poly-L-Valyl-Ribonuclease. Biochem.Biophys.Res.Comm.22, 262-267 (1966).

19. ONCLEY, J.L., ELLENBOGEN, E., GITLIN, D., GURD, F.R.N.: Protein-Protein Interaction. J.Phys.Chem.56, 85-92 (1952).

20. WAUGH, D.F.: A Fibrous Modification of Insulin. I. The Heat Precipitate of Insulin. J.Am.Chem.Soc.68, 247-250 (1946).

21. WAUGH, D.F., WILHELMSON, D.F., CUMMERFORD, S.L., SACKLER, M.L.: The Nucleation and Growth Reactions of Selected Types of Insulin Fibrils. J.Am.Chem.Soc.75, 2592-2600 (1953).

22. WAUGH, D.F.: Protein-Protein Interactions. Adv.Prot.Chem.9, 325-347 (1954).

23. DOTY, P., MYERS, F.E.: Thermodynamics of the Association of Insulin Monomers. Discussions Faraday Soc.13, 51-58 (1953).

24. CRUFT, H.J., MAURITZEN, C.M., STEDMAN, E.: The Isolation of β-histone from Calf Thymocytes and the Factors Affecting its Aggregation. Proc.Roy. Soc.B 149, 21-35 (1958).

25. SMERDON, M.J., ISENBERG, I.: The Effect of Temperature on Histone GRK Aggregation. Biochem.Biophys.Res.Comm.55, 1029-1034 (1973).

26. KLOTZ, I.M.: The Nature of Some Ion-Protein Complexes. Cold Spring Harbor Symp.Quant.Biol.14, 97-112 (1949).

27. COHN, E.J., FERRY, J.D.: Interaction of Proteins with Ions and Dipolar Ions. In: Proteins Amino Acids and Peptides (E.J.COHN, J.P.EDSALL, eds.), chap.4, pp.586-622. New York: Reinhold 1943.

28. ANDERER, F.A., LAUFFER, M.A.: unpublished experiments, 1965.

29. WARNER, R.C.: Separation of α- and β-casien. J.Am.Chem.Soc.66, 1725-1731 (1944).

30. EYRING, H.: The Activated Complex and the Absolute Rate of Chemical Reactions. Chem.Rev.17, 65-77 (1935).

31. ANSON, M.L.: Protein Denaturation and the Properties of Protein Groups. Adv.Protein Chem.2, 361-386 (1945).

32. KUNITZ, M.: The Kinetics and Thermodynamics of Reversible Denaturation of Crystalline Soybean Trypsin Inhibitor. J.Gen.Physiol.32, 241-263 (1948).

33. BROWN, D.E.S., JOHNSON, F.H., MARSLAND, D.A.: The Pressure-Temperature Relations of Bacterial Luminescence. J.Cell.Comp.Physiol.20, 151-168 (1942).

34. EYRING, H., MAGEE, J.L.: Application of the Theory of Absolute Reaction Rates to Bacterial Luminescence. J.Cell.Comp.Physiol.20, 169-177 (1942).

35. HEYMANN, E.: Dilatometric Investigations on the Heat-Denaturation of Proteins. I. Biochem.J.30, 127-131 (1936).

36. TANFORD, C.: Protein Denaturation. Adv.Protein Chem.23, 121-282 (1968).

37. JAENICKE, R.: Volume Changes in the Isoelectric Heat Aggregation of Serum Albumin. Eur.J.of Biochem.21, 110-115 (1971).

38. ROSS, P.D., SHAPIRO, J.T.: Heat of Interaction of DNA with Poly Lysine, Spermine and Mg^{++}. Biopolymers 13, 415-416 (1974).

Chapter 10

1. LAUFFER, M.A.: Protein-Protein Interaction; Endothermic Polymerization and Biological Processes. In: Proteins and Their Reactions (H.W.SCHULTZ, A.F.ANGLEMIER, eds.), pp.87-116. Westport/Conn.: Avi Publ.1964.

2. INOUÉ, S.: Organization and Function of the Mitotic Spindle. In: Primitive Motile Systems in Cell Biology, pp.549-598. New York: Academic Press 1964.

3. MARSLAND, D.: Pressure-Temperature Studies on the Mechanisms of Cell Division. In: High Pressure Effects on Cellular Processes, pp.259-312. New York: Academic Press 1970.

4. BROWN, D.E.S.: Amoebae Movement at High Hydrostatic Pressure. J.Cellular Comp.Physiol.8, 167-168 (1936).

5. MARSLAND, D.: The Effects of High Hydrostatic Pressure Upon Cell Division in Arbacia Eggs. J.Cellular Comp.Physiol.12, 57-70 (1938).

6. BROWN, D.E.S.: The Pressure Coefficient of "Viscosity" in the Eggs of *Arbacia Punctulata*. J.Cellular Comp.Physiol.5, 335-346 (1934).

7. MARSLAND, D.: The Mechanism of Cell Division; Temperature-Pressure Experiments on the Cleaving Eggs of *Arbacia Punctulata*. J.Cellular Comp. Physiol.36, 205-227 (1950).

8. MARSLAND, D., LANDAU, J.E.: The Mechanism of Cytokinesis: Temperature-Pressure Studies on the Cortical Gel System in Various Marine Eggs. J.Exptl.Zool.125, 507-539 (1954).

9. LANDAU, J.V., ZIMMERMAN, A.M., MARSLAND, D.: Temperature-Pressure Experiments on *Amoebae Proteus;* Plasmagel Structure in Relation to Form and Movement. J.Cellular Comp.Physiol.44, 211-232 (1954).

10. MARSLAND, D.: The Mechanism of Protoplasmic Streaming. The Effects of High Hydrostatic Pressure on Cyclosis in *Elodea Canadensis*. J.Cellular Comp.Physiol.13, 23-30 (1939).

11. LANDAU, J.V., MARSLAND, D., ZIMMERMAN, A.M.: The Energetics of Cell Division; Effects of Adenosine Triphosphate and Related Substances on the Furrowing Capacity of Marine Eggs (*Arbacia* and *Chaetopterus*). J.Cellular Comp.Physiol.45, 309-329 (1955).

12. ZIMMERMAN, A.M., LANDAU, J.V., MARSLAND, D.: Cell Division; A Pressure-Temperature Analysis of the Effects of Sulhydryl Reagents on the Cortical Plasma Gel Structure and Furrowing Strength of Dividing Eggs (*Arbacia* and *Cheatopterus*). J.Cellular Comp.Physiol.49, 395-435 (1957).

13. MARSLAND, D.: Mechanism of Pigment Displacement in Unicellular Chromatophores. Biol.Bull.87, 252-261 (1944).

14. MARSLAND, D., MEISNER, D.: Effects of D_2O on the Mechanism of Pigment Dispersal in the Melanocytes of *Fundulus Heteroclitus*; a Pressure-Temperature Analysis. J.Cell Physiol.70, 209-216 (1967).

15. PEASE, D.C.: Hydrostatic Pressure Effects Upon the Spindle Figure and Chromosome Movement. I. Experiments on the First Mitotic Division of Urechis Eggs. J.Morphol.69, 405-441 (1941).

16. PEASE, D.C.: Hydrostatic Effects Upon the Spindle Figure and Chromosome Movements. II. Experiments on the Meiotic Divisions of Tradescantia Pollen Mother Cells. Biol.Bull.91, 145-169 (1946).

17. INOUÉ, S.: Effect of Temperature on the Birefringence of the Mitotic Spindle. Biol.Bull.103, 316 (1952).

18. INOUÉ, S.: Motility of Cilia and the Mechanism of Mitosis. Rev.Mod.Phys. 31, 402-408 (1959).

19. ZIMMERMAN, A.M., MARSLAND, D.: Cell Division; Effects of Pressure on the Mitotic Mechanisms of Marine Eggs *(Arbacia punctulata)*. Expl.Cell Res.35, 293-302 (1964).

20. INOUÉ, S., SATO, H., TUCKER, R.W.: Heavy Water Enhancement of Mitotic Spindle Birefringence. Biol.Bull.125, 380-381 (1963).

21. MARSLAND, D., ZIMMERMAN, A.M.: Structural Stabilization of the Mitotic Apparatus by Heavy Water in the Cleaving Eggs of *Arbacia punctulata*. Exptl.Cell Res.38, 306-313 (1965).

22. MAZIA, D., DAN, K.: The Isolation and Biochemical Characterization of the Mitotic Apparatus of Dividing Cells. Proc.Natl.Acad.Sci.U.S.38, 826-838 (1952).

23. MARSLAND, D.: Antimitotic Effects of Colchicine and Hydrostatic Pressure: Sinergistic Action of the Cleaving Eggs of *Lytechinus variegatus*. J.Cell Physiol.67, 333-338 (1966).

24. MARSLAND, D., HECHT, R.: Cell Division: Combined Antimitotic Effect of Colchicine and Heavy Water on First Cleavage in the Eggs of *Arbacia punctulata*. Expl.Cell Res.51, 602-608 (1968).

25. INOUÉ, S., SATO, H., ASCHER, M.: Counteraction of Colcemid and Heavy Water on the Organization of the Mitotic Spindle. Biol.Bull.129, 409-410 (1965).

26. McMICHAEL, J., LAUFFER, M.A.: unpublished results (1972).

27. TILNEY, L.G., HIRAMOTO, Y., MARSLAND, D.: Studies on the Microtubules in Heliozoa. III. A Pressure Analysis of the Role of these Structures in the Formation and Maintenance of the Axopodia of *Actinosphaerium nucleofilum* (BARRETT). J.Cell Biol.29, 77-95 (1966).

28. HARRIS, P.: Some Structural and Functional Aspects of the Mitotic Apparatus in Sea Urchin Embryos. J.Cell Biol.14, 475-487 (1962).

29. LEDBETTER, M.C., PORTER, R.K.: A "Microtubule" in Plant Cell Fine Structure. J.Cell Biol.19, 239-250 (1963).

30. de THÉ, G.: Cytoplasmic Microtubules in Different Animal Cells. J.Cell Biol.23, 265-275 (1964).

31. ROTH, L.E., WILSON, H.J., CHAKRABORTY, J.: Anaphase Structure in Mitotic Cells Typified by Spindle Elongation. J.Ultrastructure Res.14, 460-483 (1966).

32. KANE, R.E., FORER, A.: The Mitotic Apparatus; Structural Changes after Isolation. J.Cell Biol.25 part 2, 31-39 (1965).

33. INOUÉ, S., SATO, H.: Cell Mobility by Labile Association of Molecules: The Nature of Mitotic Spindle Fibers and Their Role in Chromosome Movement. J.Gen.Physiol.50, 259-292 (1967).

34. KIEFER, B., SAKI, H., SOLARI, H.A., MAZIA, D.: The Molecular Unit of the Microtubules of the Mitotic Apparatus. J.Mol.Biol.20, 75-79 (1966).

35. STEPHENS, R.E.: The Mitotic Apparatus; Physical-Chemical Characterization of the 22 S Protein Components and its Subunits. J.Cell Biol.32, 255-275 (1967).

36. STEPHENS, R.E.: Reassociation of Microtubule Protein. J.Mol.Biol.33, 517-519 (1968).

37. STEPHENS, R.E.: Thermal Fractionation of Outer Fiber Double Microtubules into A- and B-subfiber Components; A- and B-Tubulin. J.Mol.Biol.47, 353-363 (1970).

38. TILNEY, L.G., GIBBONS, J.R.: Microtubules and Morphogenesis. The Role of Microtubules in the Development of the Primary Mesenchyme in the Sea Urchin Embryo. Biol.Bull.131, 378-379 (1966).

39. BIKLE, D., TILNEY, L.G., PORTER, K.R.: Microtubules and Pigment Migration in the Melanophores of *Fundulus Heteroclitus L.* Protoplasma 61, 322-345 (1966).

40. TILNEY, L.G., PORTER, K.R.: Studies on Microtubules in Heliozoa. I. The Fine Structure of *Actinosphaerium Nucleofilum* (BARRETT) with Particular Reference to the Axial Rod Structure. Protoplasma 60, 317-344 (1965).

41. TILNEY, L.G., PORTER, K.R.: Studies on Microtubules in Heliozoa. II. The Effect of Low Temperature on these Structures in the Formation and Maintenance of the Axopodia. J.Cell Biol.34, 327-343 (1967).

42. STEPHENS, R.E.: Microtubules. In: Subunits in Biological Systems (S.N. TIMASHEFF, G.E.FASMAN, eds.), part A, pp.355-391. New York: Dekker 1971.

43. WEISENBERG, R.C.: Microtubule Formation *in vitro* in Solutions Containing Low Calcium Concentrations. Science 177, 1104-1105 (1972).

44. BRYAN, J.: Biochemical Properties of Microtubules. Federation Proc.33, 152-157 (1974).

45. WILSON, L., BAMBURG, J.R., MIZEL, S.B., GRISHAM, L.M., CRESWELL, K.M.: Interaction of Drugs with Microtubule Proteins. Federation Proc.33, 158-166 (1974).

46. BORISY, G.G., OLMSTED, J.B., MARCUM, J.M., ALLEN, C.: Microtubule Assembly *in vitro*. Federation Proc. 33, 167-174 (1974).

47. GORTNER, R.A.: The State of Water in Colloidal and Living Systems. Trans. Faraday Soc. 26, 678-686 (1930).

Subject Index

GPSR Compliance

The European Union's (EU) General Product Safety Regulation (GPSR) is a set of rules that requires consumer products to be safe and our obligations to ensure this.

If you have any concerns about our products, you can contact us on ProductSafety@springernature.com

In case Publisher is established outside the EU, the EU authorized representative is:

Springer Nature Customer Service Center GmbH
Europaplatz 3
69115 Heidelberg, Germany

Batch number: 10584556

Printed by Printforce, the Netherlands